WORKBOOK ON SYSTEMS ANALYSIS AND DESIGN

With Best Compliments
From
Cybertech Systems and Software Ltd.
B-65, J.B. Sawant Marg, Rd. No. 21,
Wagle Estate, Thane - 400 604.
Tel. : 91-22-5834643/44/45
Fax : 91 22-5832574

WORKBOOK ON SYSTEMS ANALYSIS AND DESIGN

SECOND EDITION

Vinod Kumar Garg
S. Srinivasan
CyberTech Systems & Software Ltd.
Mumbai

Prentice-Hall of India *Private Limited*
New Delhi-110 001
2000

Rs. 125.00

WORKBOOK ON SYSTEMS ANALYSIS AND DESIGN, 2nd Ed.
by Vinod Kumar Garg and S. Srinivasan

© 2000 by Prentice-Hall of India Private Limited, New Delhi. All rights reserved. No part of this book may be reproduced in any form, by mimeograph or any other means, without permission in writing from the publisher.

ISBN-81-203-1724-6

The export rights of this book are vested solely with the publisher.

Third Printing (Second Edition) ... ... **August, 2000**

Published by Asoke K. Ghosh, Prentice-Hall of India Private Limited, M-97, Connaught Circus, New Delhi-110001 and Printed by Meenakshi Printers, Delhi-110006.

To the all pervading
Lord Krishna
निमित्तमात्रं भव सव्यसाचिन्

Contents

Preface ix
Preface to the First Edition xi
Acknowledgments xiii
Chapter Summary xv

1. **Introduction**   1–5
   1.1 Systems Development Life Cycle *1*
   1.2 Components of Information System *3*
   1.3 Modeling in Systems Development *3*
   1.4 Application Modeling Techniques *4*
   1.5 Summary *5*

2. **Application Modeling**   6–42
   2.1 Context Diagram *6*
   2.2 Data Flow Diagram *18*
   2.3 E-R Diagrams *34*

3. **Database Design**   43–66
   3.1 Data Dictionary *43*
   3.2 Logical Data Analysis *49*

4. **Input-Output Design**   67–86
   4.1 Input Form Design *67*
   4.2 Input Screen Design *74*
   4.3 Menu Design *81*
   4.4 Output Design *83*

5. **Program Design**   87–120
   5.1 Introduction to Structured Program Design *87*
   5.2 Structure Charts *89*
   5.3 Process Specifications *94*
       5.3.1 Process Narratives *94*
       5.3.2 Decision Trees *95*
       5.3.3 Decision Tables *98*
       5.3.4 Structured English *102*
   5.4 Program Coding Standards *107*
   5.5 Software Testing Techniques *110*

6. **Case Study** — 121–142
   6.1 Overview *121*
   6.2 Analysis *126*
   6.3 Case Study #2 *139*

7. **Object-Oriented Analysis and Design** — 143–173
   7.1 SSAD vs OOAD *143*
   7.2 Concepts and Terminology *148*
   7.3 Object-Oriented Themes *154*
   7.4 How to Identify Objects and Group them into Classes and Super Classes *158*

8. **Question Bank** — 174–208
   Exercises *174*
   Multiple Choice Questions *202*
   True/False Questions *206*

Appendix A — 209–215
   Structured Methodology Elements *209*

Appendix B — 217–229
   Web Case Study *217*
   Function Specification *218*
   Design Specification *221*
   HTML vs ASP *223*

Glossary — 231–234

Index — 235–236

Preface

In its second edition, this workbook continues its step-by-step approach to the concepts of systems analysis and design to enable students understand and implement the complete flow of the systems design and processes.

The idea of this second edition has germinated from the practical problems that we have faced as teachers. We have also noticed that students are often stumped when faced with real world computer applications. We have therefore focussed on various practice exercises simulating a small real world system and tried to impart practical knowledge on the subject. We have generally tried to act as a beacon that helps the students navigate through the finer aspects of the subject for a thorough understanding of the same.

Today, all the developments in systems design are web centric based on object-oriented technologies. Object-oriented design is an approach that is well suited to a client-server hardware and software platform which is the most popular architecture for running applications. Thus, new chapters have been added for object-oriented analysis and design and web application development. Moreover, the chapters of the first edition have been revised and expanded.

We believe that the information contained in this edition will go a long way in the career advancement of students in today's fast-paced and breathtaking arena of IT. Helpful suggestions and comments for any new features and topics for improvement of the text will be greatly appreciated.

Vinod Kumar Garg
S. Srinivasan

Preface to the First Edition

Although there are many textbooks which illustrate how to develop, test and implement software products based on various methodologies, the need of the hour is to provide the students with a workbook that would help them understand the concepts of systems analysis and design. Courses on systems development, normally taught in computer science in many colleges and institutions, follow a sequence of instructions in structured programming. But when it comes to implementing them in the projects, students find it difficult to pick the threads so as to establish the complete flow of the system design processes. This book will help them just do that.

Providing the opportunity to practise skills is exciting to the learner in any systems development course. This works best in a project or case study environment, or through a project simulating a small real world system.

While teaching this course in various institutions, we have ourselves faced problems to infuse in the students the concepts of systems analysis and design through applications that are implemented in live environments. Keeping this in mind, we focussed on providing various examples, exercises and practices so as to enable the discerning students to get adequate understanding of the concepts underlying the subject of systems analysis and design. We therefore feel that the book should also be helpful and handy to instructors in conducting practice sessions and assisting students to gain a much broader view of the subject.

In this workbook, the examples are introduced in the sequence in which they would be needed during systems analysis and design. The book first outlines the steps followed in analysis and design and then illustrates with examples followed by exercises. At the end of each chapter, practice exercise is given for students to solve them. The last two chapters present case study and question bank respectively. The question bank gives diverse types of software systems which can be automated.

Any constructive suggestions to include new features/topics for improvement of the text will be greatly appreciated.

Vinod Kumar Garg
S. Srinivasan

Acknowledgments

Motivation for writing a book of this type came from our experience of teaching where we identified problems generally encountered by students to get a grasp of the subject. We thank all those who directly or indirectly helped us complete this project. We are grateful to N.K. Venkitakrishnan and Lata Chablani of CyberTech Systems & Software Ltd. for their assistance in preparing the manuscript.

Our sincere gratitude goes to Sarita Garg, Neeti Garg, Vikram Garg, Vijaya Srinivasan and Krishna Srinivasan for all the encouragement, patience and special care shown by them during the preparation of the manuscript.

We are also grateful to our parents Dr. R.P. Garg, Kailashi Garg, V. Sitaraman, Padma Sitaraman for their constant blessings and guidance.

Finally, we wish to express our sincere thanks to the publishers, Prentice-Hall of India for the meticulous processing of the manuscript, both during editorial and production stages.

Vinod Kumar Garg
S. Srinivasan

Chapter Summary

Chapter 1   Introduction

This chapter briefly outlines the nature and components of an information system. It also highlights the role of modeling in systems development.

Chapter 2   Application Modeling

This chapter describes three most important techniques for application modeling, namely, context diagram, data flow diagram and E-R diagram. It also provides practical hints on how to use these techniques in actual practice.

Chapter 3   Database Design

This chapter provides both top-down and bottom-up approaches to design a database for any application under consideration. It also shows step-by-step approach to carry out logical data analysis and how to build a data dictionary, given a high level business analysis.

Chapter 4   Input-Output Design

This chapter provides very detailed guidelines to develop the most important features of today's application, namely, user interface.

Chapter 5   Program Design

This chapter explains in a step-by-step manner how to convert a high level design depicted by data flow diagram into structure charts and supporting process specifications. Various process specification tools, namely, decision trees, decision tables and structured English are explained with special emphasis on how to use these tools in a given situation. This chapter also covers very important topics on program coding standards and testing techniques and how to relate them to program design.

Chapter 6   Case Study

A comprehensive case study is discussed here to illustrate how the various application modeling tools could be applied in an integrated way to a real life situation. All the chapters are supported by a three-tier structure of examples, exercises and practice to provide an incremental, progressive and spiral learning approach.

Chapter 7   Object-Oriented Analysis and Design

Object-oriented design is the approach for today's design and development of application. This chapter explains the difference in SSAD and OOAD techniques. Object-oriented design is an approach that uses objects, classes and messages to model processes in real world. This approach is well suited to a client-server hardware and software platform, which is the most popular architecture for running applications.

Chapter 8   Question Bank

This final chapter is a rare contribution where carefully selected exercises and questions on various concepts are provided at one place. This should be of tremendous help to instructors and students preparing for various examinations and to those who really want to cement the knowledge gained through earlier chapters.

Appendix and Glossary

Appendix—A is top-town pictorial representation of system analysis and design process, under the heading "Structured Methodology Elements". This will help one to understand all the processes in system analysis and design activity at one go.

Appendix—B is a case study to develop web applications. This will be of great help to understand how designing web based applications is different from normal applications.

Also appended is glossary of terms and their meanings, which can be referred to get the basic definitions.

1

Introduction

1.1 SYSTEMS DEVELOPMENT LIFE CYCLE

Good management of software projects is vital. Many systems arrive late, cost more than was budgeted earlier and do not meet user requirements. Systems development life cycle is concerned with the detail management of all the components that go into the development of a new system.

The systems development life cycle was developed by the National Computing Centre in the 1960s to add discipline to many organizational approaches to systems development. It is a model of how systems should be designed and developed.

Over the years, lot of methodologies have evolved to support this life cycle approach of systems development. **A methodology is a collection of procedures, techniques, tools and documentation aids** that are designed to help systems developers in their effort to implement a new system.

The quality of the systems you design and implement depends on the adequacy of their model for prescription and description of system development activities. The life cycle approach arose out of the early efforts to apply project management techniques to the system development process. Table below explains the phases in order of systems development life cycle.

Phase	Explanation
Feasibility study	Applying cost-benefit criteria to the proposed application. This will involve problem definition, entry and feasibility assessment.

Information analysis	Studying user information requirements.
	This will involve analysis of the existing system, the various interfaces to other systems, etc.
	Terms of reference for the system.
	    — Boundaries of the system to be examined
	    — Proposed objectives of the new system

(Contd.)

2 *Workbook on Systems Analysis and Design*

Phase	Explanation
	— Resource/Organizational constraints — Strategy for developing the system Formal approval leading to freezing the scope of system.
System design	Designing files and information processing functions to be performed in the system. This will led to data flow diagram, normalized data structures, program specifications.
Procedures and forms development	Designing and documenting procedures and forms for system users.
Program development	Designing, coding, compiling, testing and documenting programs.
Acceptance testing	Prepare and carry out final system test and formal approval and acceptance by users and management.
Conversions	Changeover from old to new systems.
Operation and maintenance	Ongoing running of system and subsequent Changes/ Improvement
Post audit	Periodic review of the system.

Responsibilities during SDLC

Stages	Users	Enterprise	System Analyst
Inception	Initiate Study and need for applications Describe existing procedures	As sponsor approve and set objectives for application	Study and listen to user requirements, respond to questions, device alternatives, Work out rough cut plan and estimates
Feasibility Study	Evaluate existing applications/systems and proposed alternatives	Review and choose alternatives from the estimates	Evaluate alternatives using agreed upon alternatives
System Analysis	Describe existing systems, Collect data of inputs, processes and output.	Provide resources and attend steering committee reviews.	Analyze and Collate data, document findings and work out the scope of the system
Design	Process logic plan for data migration, design manual procedures and remain aware of design changes	Encourage user friendly design, review design and recommend business process changes, plan impact.	Present alternatives and trade-off to users for their decision.

Stages	Users	Enterprise	System Analyst
Construction	Review program specifications	Understand high-level logic and key features Monitor and provide extra resources if any	Translate logical design into process specification for coding. Develop technical plan for data migration.
Testing	Generate test data and evaluate test results	Review	Carry out Unit test, Integration test and System test.
Conversion and Installation	Conduct end user training, Phase conversion	Attend user sessions and commitment of its usage	Perform conversion-processing tasks, User manual.
Go Live	Monitor system usage and quality of process and outputs. Suggest modifications and enhancements.	Monitor and ROI (Return on Investment)	Respond to enhancement request, Suggest improvements.

1.2 COMPONENTS OF INFORMATION SYSTEM

An information system integrates five components—**People, Procedures, Data, Software and Hardware**—to produce information by accessing and processing data of these. Data is the centre of the system. Both hardware and people are sources of activity. Both software and procedures are sets of instructions; software instructs hardware, whereas procedures instruct people. However, during systems design and development equal efforts need to be spent on each of these components.

Also when a system is computerized, many activities formerly performed by people following physical procedures are instead done by hardware, executing applications software. Nonetheless, both people and procedures are still very important components of the computerized system. And it is the homogeneous interfacing between these, which the computerized application needs to handle.

1.3 MODELING IN SYSTEMS DEVELOPMENT

Systems analysis is the key component of the first two phases of the systems development life cycle.

In **Phase I,** systems analysis techniques are put to use, to help build an understanding of the existing systems, of the business need and of the potential solutions to meet that need. **Phase II** of systems analysis is used to further this understanding and to produce specifications for a new system that will meet user needs and requirements.

An important issue in systems design is the question of centralization or decentralization. **Centralization** allows the use of a single set of information and affords better security and control. This has been the way the old **legacy systems** have been developed and implemented.

But with the advent of client server technology, the information is being assembled and processed across distributed platforms and environment. Designing and development of systems have also become more complicated. This has necessitated the use of formal modeling techniques in analysis and design phase to identify and incorporate all the components.

1.4 APPLICATION MODELING TECHNIQUES

Systems analysis can be a complex and confusing work. The analyst should therefore be able to deal with large amounts of highly detailed and often conflicting information. The analyst needs a way to organize the information, determine where there are gaps in understanding, and identify areas of conflicting or redundant operations. **Modeling techniques** provide this mechanism.

Some of these techniques which are used in systems development are:

- Context diagrams
- Data flow diagrams
- E-R diagrams

The key challenge of analysis is to achieve effective communication among users, analysts, and systems designers, and these tools serve the purpose. The context diagram defines the scope of the system. Models based on data flow diagram are very effective during the analysis activities of the development process. They emphasize the flow and processing, or transformation of data within the system.

Data flow diagrams alone are not sufficient to model a system. The data dictionary is used to maintain definitions of the data in normalized form that constitute the data flows and data stores in a data flow diagram. Finally, process descriptions are used to document intermediate and low-level process, which is used as implementable units at the time of development.

Context Diagram

In keeping with the top-down approach to requirement determination, the first graphic that is produced using structured technique is the **context diagram**. It gives a broad overview of the information system environment, including data flows into and out of the system.

Data Flow Diagram

The **data flow diagrams (DFD),** the next tool in this top-down approach, moves from general requirements to more specific requirements, illustrating the processes, movement, and storage of data in the system. In DFD, processes are first identified and then the data flow between the processes are isolated and derived. Thus **processes are the focal point of the DFD**. Data flow diagrams themselves stress the flow and transformation of data within a system.

Entity-Relationship (E-R) Diagram

The next tool is the **entity-relationship (E-R) diagram**. In this the entities (data objects) are isolated and the relationships between them are defined. E-R diagrams stress upon data and how they are organized and are accessed. Thus **data is the focal point of E-R** diagrams.

1.5 SUMMARY

Our approach to requirement determination is the top-down approach. The top-down approach begins with the organization-wide picture by way of context diagram. After the scope is defined and the areas of application development are determined, we progress to more concrete levels of the detail of each process and the information flow between these processes. This results in data flow diagram as output.

Data flow diagrams alone are not sufficient to model a system. Before the system is rolled for actual development, the data dictionary, used to maintain the data that constitutes the data flow and data stores, are normalized. This results in E-R diagram.

2

Application Modeling

A number of different modeling techniques are used in systems development. These techniques, because of their simplicity and clarity, are very effective for enhancing communications between users and systems designers. Structured techniques such as context diagrams, data flow diagrams, and E-R diagrams are designed to reduce the amount of time spent in reacting to change requests during and after systems development.

2.1 CONTEXT DIAGRAM

A context diagram is a structured graphical tool for identifying the organization's functional areas and the processes that are performed within and between the organization and the outside world.

Context diagrams serve three important purposes whereby you can determine preliminary requirements:

1. Context diagrams support a data-oriented approach to system design,
2. help you investigate the output and process requirements of the organization, and
3. help you define the boundaries of the proposed system.

Symbols Used in Context Diagram and Data Flow Diagram

Three symbols are used in a context diagram to represent external entities, processes and data flows, as shown in Figure 2.1(a). The fourth symbol, the 'data stores' is used in data flow diagram.

External entity

A rectangle indicates any entity external to the system being modeled. The entity can be a class of people, an organization, or even another system.

The function of the external entity is to, supply data to, or receive data from the system. They have absolutely no interest in how to transform the data.

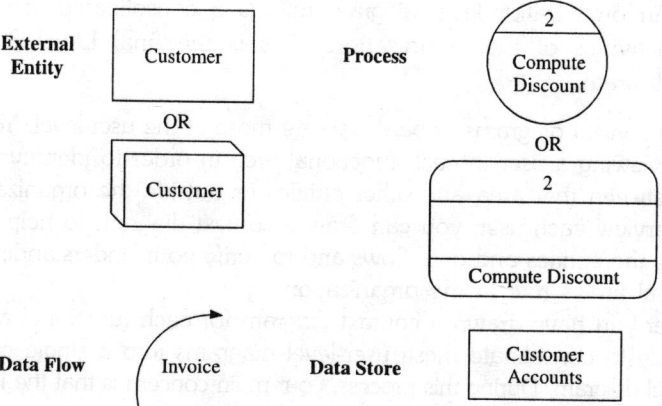

Figure 2.1(a) Data flow diagram symbols.

Process

Bubbles or circles are used to indicate where incoming data flows are processed and then transformed into outgoing data flows. This can also be represented by rectangles as shown in Figure 2.1(a). The processes are numbered and named to indicate the occurrence in the system flow.

Data flow

Arrows marking the movement of data through the system indicate data flows. It's the pipeline carrying packets of data from an identified point of origin to a specific destination. In plain terms, data flow is a transfer of data between two entities.

> **Example**
> A customer (external entity) places an order (data flow) with the order-processing department (process).

Data store

Open rectangles are used to identify holding points for data. Data stores denotes data in rest. For example, general ledger, account details, item details, etc. are some of the data stores.

Levels of Context Diagrams

There are three levels of context diagrams:

1. A user-level diagram describes one functional area's operational activity. This is Level 1 context diagram.
2. A combined user-level diagram provides an overall view of the activities of related user groups. This leads to Level 2 context diagram.

8 *Workbook on Systems Analysis and Design*

3. An organization-level diagram reflects a consolidated view of the activities of the organization. This is the final Level 3 context diagram.

The first context diagrams to be drawn are those at the user-level. You begin by interviewing a user in each functional area in order to identify the data flows between that area and other entities or outside the organization. As you interview each user, you can draw a context diagram to help the user visualize the entities and data flows and to verify your understanding of the functional area's role in the organization.

After you have drawn a context diagram for each functional area, you are ready to consolidate these user-level diagrams into a single combined user-level diagram. During this process, your main concern is that the individual user-level views are consistent with one another.

Once consistency is achieved, you can simplify the combined user-level context diagram by focusing on only those entities and data flows affected by the proposed system. So your next step is to draw a system boundary line around all the internal entities that will use the proposed system.

Example

The company, CAT Logistics is into warehousing business. The consultant from your organization has been invited to do the initial study and evolve the scope of the system. The system proposed is an "Order Tracking System". The departments involved are Order Processing, Warehouse Shipping, Warehouse Receiving and Purchase departments. The company has furnished from each department a list of the players who fall into the scope of the study.

Players
Srinivasan — Consultant
Lata — Order Processing Department
Reshmi — Warehouse Shipping
Bhushan — Purchasing
Debie — Warehouse Receiving

Discussion
During Srinivasan's first interview, Lata described how the Order Processing department should interact with the customer (an external entity) and Warehouse Shipping (an internal entity) to process an order. But first, the customer places an order with the Order Processing department.

The Order Processing department then requests the Warehouse Shipping to ship the product to the customer and notify Order Processing department about the shipment. The Order Processing department then prepares an invoice and sends it to the customer.

Further, Srinivasan met Reshmi from the Warehouse Shipping department. She confirmed that the Warehouse Shipping department was responsible for inventory management. According to her, when stock is low, Shipping notifies Order Processing which keeps a copy and sends another copy to Purchasing. Purchasing places and order with the product supplier. When the product is received, Warehouse Receiving confirms receipt to Purchasing.

Thereafter Srinivasan met Bhushan of Purchasing, who explained that Purchasing notifies Order Processing when a product is ordered and again when the product is received. Purchasing awaits confirmation of product receipt before processing and paying invoices from the supplier.

Finally Srinivasan met Debie of Warehouse Receiving, and confirmed the details he had collected. The above information was compiled by Srinivasan, and he used the same to build the context diagram for the proposed system.

This is illustrated through series of levels/steps and is discussed in Figure 2.1(b) through Figure 2.1(e).

Level 1 (Step 1)

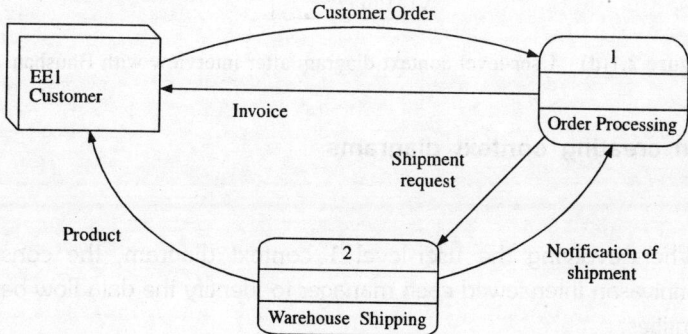

Figure 2.1(b) User-level context diagrams after interview with Lata.

Level 1 (Step 2)

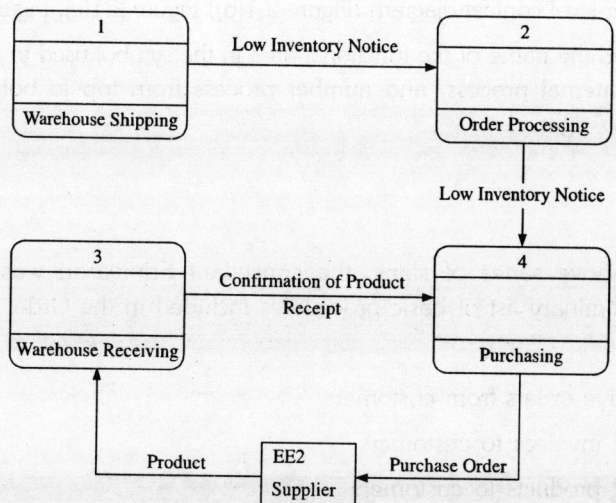

Figure 2.1(c) User-level context diagram after interview with Reshmi.

Level 1 (Step 3)

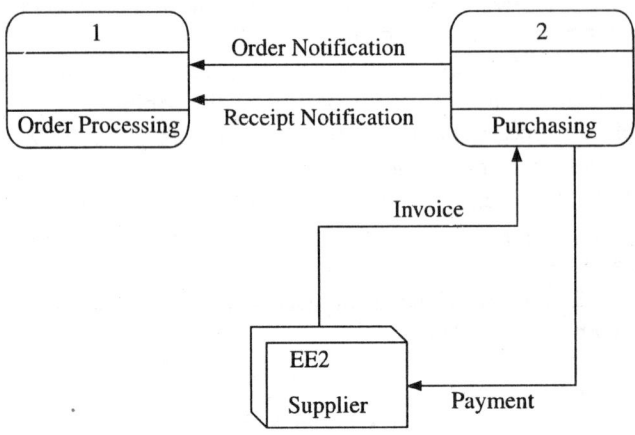

Figure 2.1(d) User-level context diagram after interview with Bhushan.

Steps in creating context diagrams

Level 1

1. When evolving the user-level 1 context diagram, the consultant Srinivasan interviewed each manager to identify the data flow between entities.
2. Identified the external entities such as customer and supplier.
3. Next he documented the information gathered from each manager in a user-level context diagram (Figure 2.1(b)) Figure 2.1(c), Figure 2.1(d).
4. Placed the name of the functional area in the symbol used to represent the internal process, and number process from top to bottom and left to right.

Level 2

From the above series of steps, the consultant Srinivasan was able to make a preliminary list of basic procedures included in the Order Tracking system.

1. Receive orders from customers
2. Send invoices to customers
3. Send products to customers
4. Send orders to suppliers

5. Receive invoices from suppliers
6. Send payments to suppliers
7. Receive products from suppliers.

This results in a combined user-level context diagram as shown in Figure 2.1(e).

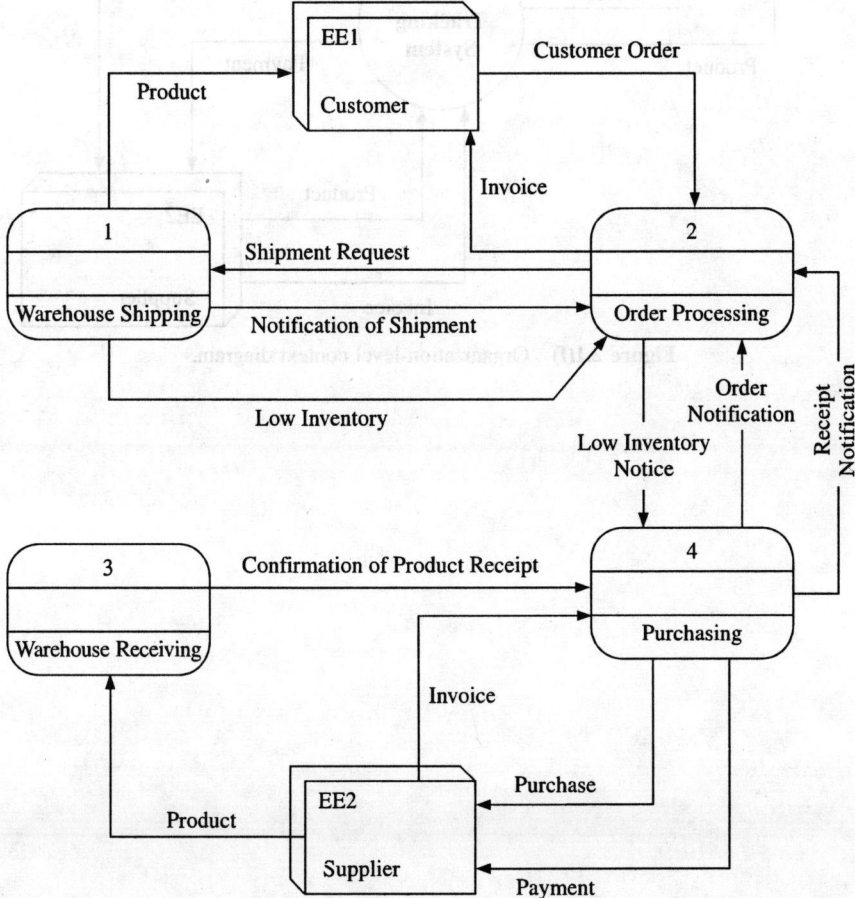

Figure 2.1(e) Combined user-level context diagram for Order Tracking system.

Level 3

Based on the steps of Level 1 and Level 2, an organization-level context diagram was drawn out to identify the general organization requirements for the Order Tracking system as shown in Figure 2.1(f).

12 *Workbook on Systems Analysis and Design*

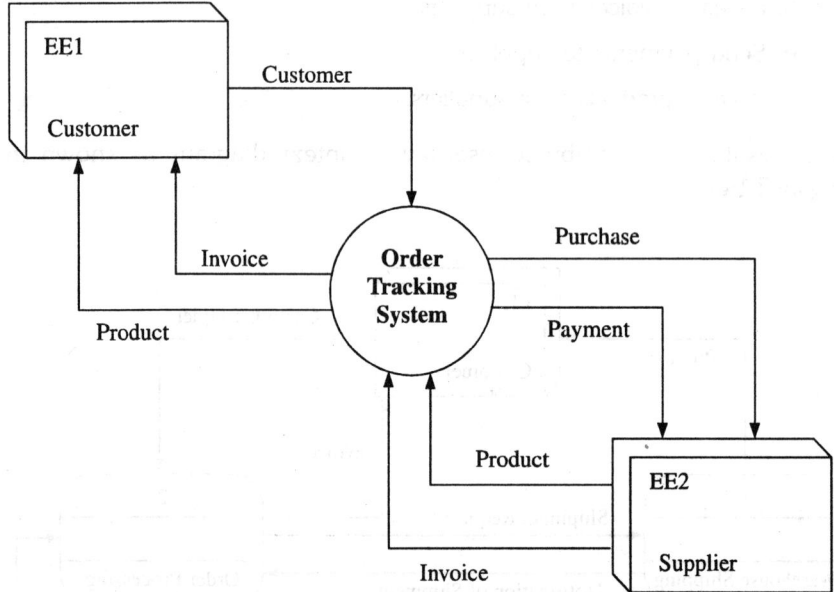

Figure 2.1(f) Organization-level context diagram.

Application Modeling

Conclusion

From Figure 2.1(f) it is clear that customer places order for the product and receives invoice and the product. Similarly the system, after receiving the request for product from the customer, places a purchase order with the supplier and thereafter receives the product along with invoice. **But the real power here is that the scope of the entire system can be understood at a glance. Also Input and Output, which are going to become interface, are identified.**

Summary

The key challenge of analysis is to achieve effective communication among users, analysts, and system designers. Effective communication has two parts: **presentation** and **understanding**. These two parts help you identify those factors that are critical to the organization success. The understanding of this is what translates into context diagram. The context diagram is a structured tool that focuses more specifically on system requirements and boundaries.

Drawing a user-level context diagram for each functional area gives you a chance to learn about the role each plays in accomplishing the organization functions and objectives. Consolidating the user-level diagrams to form one combined user-level context diagram provides a consistent, verified view of organization inputs and outputs to each process.

Exercise

Draw the context diagram based on the following narrative text.

The customer sends a list of items required, which is processed by the Customer Handling department. A copy of the list is sent to the stores. Based on the item price, an estimated value of goods is prepared and sent to the client. At the end of the month a consolidated list of customer requests is prepared and sent to manager of the Sales department.

Solution to Exercise

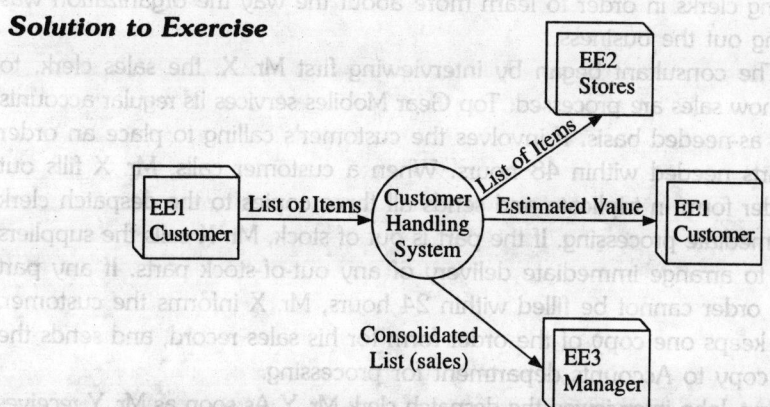

Figure 2.1(g) Context diagram for Retail store.

Practice for Section 2.1

#1

Figure 2.1(h), illustrates a context diagram for student registration system. Try to identify the external entities, process and data flows for the system.

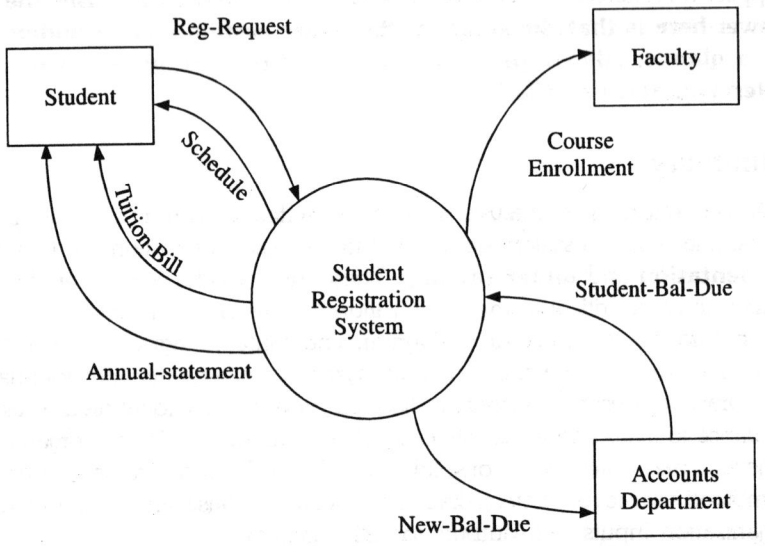

Figure 2.1(h) Context diagram for student registration system.

#2

The following is the information, a system analyst has gathered based on his interview with the various users at "Top Gear Mobiles". The company is into wholesale distribution of automotive parts. The company wants to install an Inventory and Sales Management system to help keep track of sales and manage its inventory better. The consultant's first step to investigate the requirements of the proposed system was to interview the sales and shipping clerks in order to learn more about the way the organization was carrying out the business.

"The consultant began by interviewing first Mr. X, the sales clerk, to learn how sales are processed. Top Gear Mobiles services its regular accounts on an as-needed basis. It involves the customer's calling to place an order for parts needed within 48 hours. When a customer calls, Mr. X fills out an order form in triplicate and sends all three copies to the despatch clerk for immediate processing. If the part is out of stock, Mr. X calls the suppliers to try to arrange immediate delivery of any out-of-stock parts. If any part of the order cannot be filled within 24 hours, Mr. X informs the customer. Mr. X keeps one copy of the order form for his sales record, and sends the other copy to Accounts department for processing.

Next John interviewed the despatch clerk Mr. Y. As soon as Mr. Y receives an order form from the Sales department, he fills the order noting any items not in stock on the order form. Then Mr. Y places one copy of the order form in the box with the parts and puts the box on the to-be-delivered dock, where the delivery truck driver picks it up for his next round of deliveries. Mr. Y returns the other two copies of the order form to Mr X.

Based on the above description:

(a) **Prepare a user-level context diagram for consultant interview with Mr. Y.**

(b) **Prepare a user-level context diagram for consultant interview with Mr. X.**

(c) **Consolidate the diagrams in (a) and (b) to create a combined user-level context diagram.**

(d) **Prepare an organization-level context diagram including all appropriate details.**

#3

List the errors in the user-level context diagram in Figure 2.1(i). Draw a corrected version of the diagram.

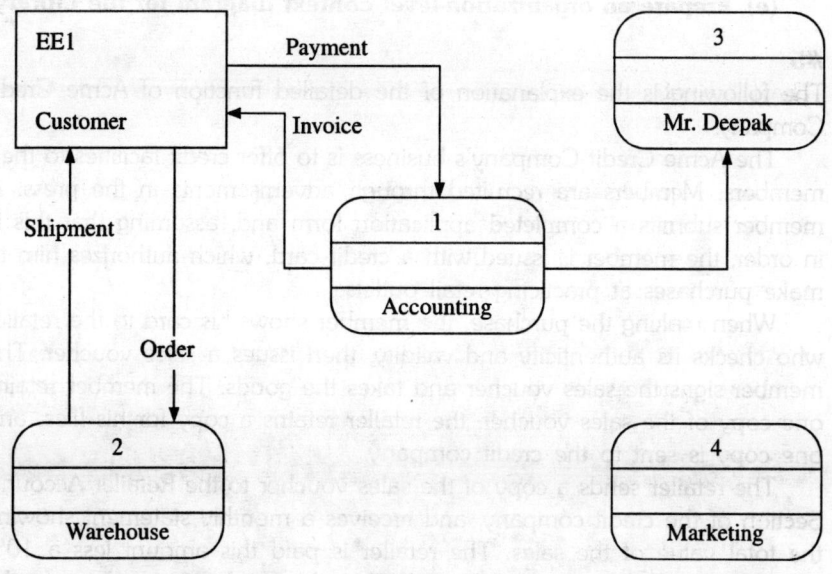

Figure 2.1(i)

#4

"Books on Hire" is a huge circulating library of books and periodicals. The following is the operation of the library. There are three departments—Library, Purchase and Accounts. Members borrow books on regular basis. If a new member joins, he has to complete a membership form and pay the fees to get himself registered. If a book asked by a member is out on loan, the member may complete a reservation form to reserve the book when it is returned. If a certain book is not available, a member may complete a suggestion form to suggest that the Library purchase the book. The librarian sends the list of books and periodicals to be purchased on regular basis to Purchase department. The Purchase department gets the quotations of the books from various suppliers and places order with them. After procuring, the Purchase department sends the books to the Library and the invoices to Accounts department. The Accounts department settles the bills of the suppliers.

Based on the above description:

(a) **Prepare a user-level context diagram from the information you have collected from the librarian.**
(b) **Prepare a user-level context diagram from the information you have collected of Purchase department.**
(c) **Prepare a user-level context diagram from the information you have collected of Accounts department.**
(d) **Prepare a user-level context diagram that summarizes the information of Accounts department.**
(e) **Prepare an organization-level context diagram for the Library.**

#5

The following is the explanation of the detailed function of Acme Credit Company.

The Acme Credit Company's business is to offer credit facilities to their members. Members are recruited through advertisements in the press. A member submits a completed application form and, assuming that this is in order, the member is issued with a credit card, which authorizes him to make purchases at practicing retail outlets.

When making the purchase, the member shows his card to the retailer who checks its authenticity and validity, then issues a sales voucher. The member signs the sales voucher and takes the goods. The member retains one copy of the sales voucher, the retailer retains a copy for his files, and one copy is sent to the credit company.

The retailer sends a copy of the sales voucher to the Retailer Accounts Section of the credit company, and receives a monthly statement showing the total value of the sales. The retailer is paid this amount less a 10% service charge. This copy of the sales voucher is then passed on to the Member Accounts Section. Sales vouchers sometimes get lost because of the volume of paperwork going around inside the company.

The Member Accounts Section send a monthly statement of account to the member, against which the member must pay 5% of the total due, but if this is less than £5, he should pay the full amount. The member may, if he wishes, pay the full outstanding amount. The payment must be sent to the credit company within 20 days of the date of the statement. If the full outstanding amount is not paid, then interest is charged on the full amount and the balance is carried forward to next month. If the member makes no payment, then a reminder letter is sent. A legal letter follows this if no payment is received in response to a second monthly statement. If a member has no outstanding balance then no statement is sent.

Based on the above description:

(a) **Make a list of the external entities which send or receive information, documents or physical resources.**

(b) **For the Acme Credit Company, make a list of the documents and physical resources—such as materials and goods which flow within the system.**

(c) **Prepare an organization-level context diagram.**

#6

Prime Placements is a recruitment agency which currently uses a manual system for recording job vacancies and applicant details for various companies/organizations. Consideration is being given to computerizing the system. You have been asked to make the analysis of the current operations. The details of the operations are as follows:

Currently the Prime Placement has assigned clerks, whose responsibility is to answer telephone queries about suitable job vacancies. Job seekers who want to work in central suburbs between Kalyan and Dadar, come to the office to ask if there are any suitable vacancies for a shorthand typist with clerical experience.

Similarly various organizations also give their requirements to Prime Placements. The agency scans their information, shortlists the candidates meeting the criteria, and after initial screening, sends the candidates to the company for final screening.

Based on the above description:

(a) **Make a list of the external entities which send or receive information to/from the agency.**

(b) **For the Prime Placements, make a list of the documents and information flowing within the system.**

(c) **Prepare an organization-level context diagram.**

2.2 DATA FLOW DIAGRAM

A DFD is a pictorial representation of the path which data takes from its initial interaction with the system until it completes any interaction. The diagram will describe the logical data flows without detailing the movements of any physical items. The DFD also gives insight into the data that is used in the system, i.e. who actually uses it and where it is temporarily or permanently stored.

A data flow diagram (DFD) does not show a sequence of steps. A DFD only shows what the different processes in a system are and what data flows between them.

Rules for Data Flow Diagram

Broad guidelines on how to draw a data flow diagram:

- Overall
    1. Fix the scope of the system by means of context diagram.
    2. Organize the DFD so that the main sequence of actions reads left to right and top to bottom.
    3. Identify all inputs and outputs to the system.
- Processes
    4. Identify and label each process internal to the system with rounded circles.
    5. A process is required for all data transformations and transfers. Therefore, never connect a data store to a data source or destinations or another data store with just a data flow arrow.
    6. Do not indicate hardware and ignore control information.
    7. Make sure the names of the processes accurately convey everything the process is doing.
    8. There must not be unnamed process.
- External Entities
    9. Indicate external sources and destinations of data, with squares.
    10. Number each occurrence of repeated external entities.
- Data Flows
    11. Identify all data flows for each process step, except simple record retrievals.
    12. Label data flows on each arrow.
    13. Use data flow arrows to indicate data movement.
    14. There cannot be unnamed data flow.
    15. A data flow cannot connect two External entities.

> Data Stores
>> 16. Do not indicate file types for data stores.
>> 17. Draw data flows into data stores only if the data store will be changed.
>> 18. To access data store, first data must be written into it.

Exercises

#1
Once the scope of the system is defined, External entities can be identified. (True/False)

#2
The scope of the system is fixed when:

(a) The External entities are defined.

(b) External entities and Inputs and Outputs have been identified.

Choose the correct option.

#3
An External entity can access a data store directly, without an intervening process. (True/False)

#4
All the data, which flows into a process, must be reflected in the output of the process. (True/False)

#5
All the data needed to produce the output must be available to the process. (True/False)

#6
Whenever an account holder wants to withdraw some cash he presents a cheque or withdrawal slip. The account is checked for the requisite balance. If the necessary balance exists, the cash is paid and the account updated. The account holder's passbook is updated, indicating the withdrawal.

(i) Identify the External Entity for the above system description.

(ii) Identify the data flows from and to the External Entity.

(iii) In the list of inputs/outputs, will cash and passbook need to be shown?

Solutions to Exercises

#1 True.
First the scope of the system is defined and then whatever lies outside the scope but interacts with the system, would be considered as an external entity.

#2 (b)

20 Workbook on Systems Analysis and Design

#3 False.
An external entity cannot access or update a data store directly.

#4 True.

#5 True.
A process cannot generate data on its own. It only converts some input to output.

#6 (i) Account holder.

  (ii) (a) From the external entity—Cheque, withdrawal-slip.

  (b) To the external entity—Update Passbook.

  (iii) No.

  In DFD, flow of data is shown and not movements of physical thing like Cash or Passbook.

Convert Context Diagram to Data Flow Diagram

Preparing a context diagram is a preliminary step in creating a data flow diagram (DFD). Based on context diagram, data flow diagrams identify the major data flows within the system boundaries, the process and the data storage (Figures 2.2(a) and 2.2(b)).

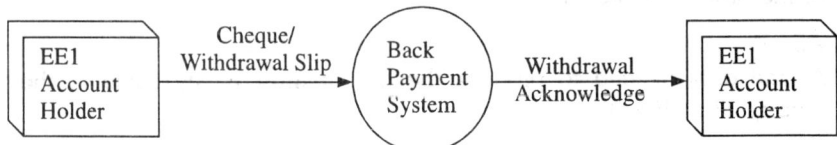

Figure 2.2(a) Context diagram.

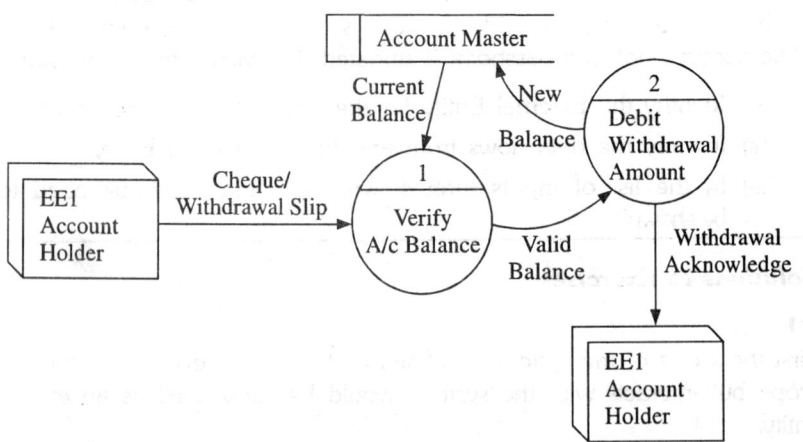

Figure 2.2(b) Data flow diagram.

Application Modeling

> **Example**
>
> Consider a system in bank whereby the account holders get their withdrawals effected. Whenever an account holder wants to withdraw some cash, he presents a cheque or withdrawal slip. The account is checked for appropriate balance. If balance exits, the cash is paid and the account is updated.

Figure 2.2(a) shows the context diagram, and Figure 2.2(b) shows the corresponding data flow diagram.

In the above DFD, we could have added another process; Verify Cheque Number, to check if the cheque submitted by the account holder is valid. This will be validated against another data store; Cheque Book Issued.

Showing data flows correctly

It's very important to identify between data flows and control flow. In fact, control signals are not to be depicted as data flows. If the information in a data flow is used only to start a process and is not being modified by the process, then it is a control flow. **Thus data flows arrows link all the processes to data stores in DFD.**

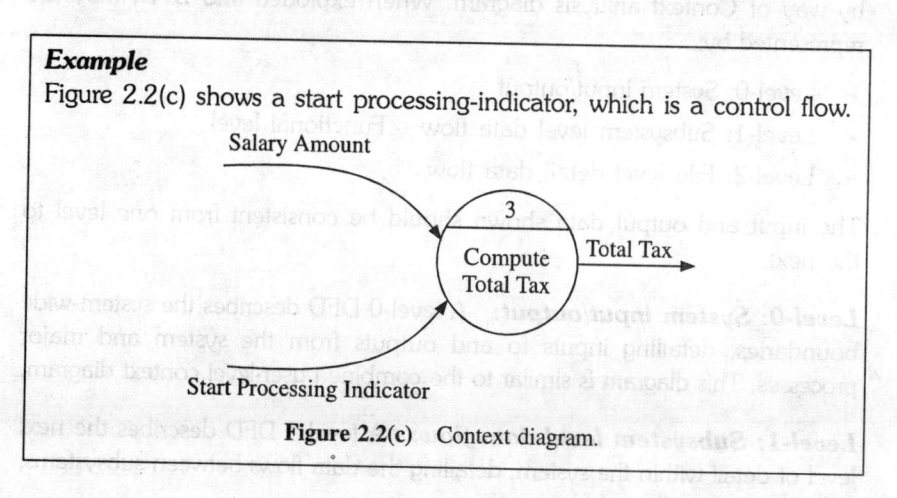

> **Example**
>
> Figure 2.2(c) shows a start processing-indicator, which is a control flow.
>
> Salary Amount → 3 Compute Total Tax → Total Tax
>
> Start Processing Indicator
>
> **Figure 2.2(c)** Context diagram.

Another way to determine whether a data flow is in fact a data flow or a control flow is to try and identify the information flowing. If there is none, then it is most likely to be a control flow.

> **Exercise**
>
> Consider a situation where cards are accepted one after the other from a department and classified as correct or incorrect cards as depicted in Figure 2.2(d). Identify control flow(s).

Figure 2.2(d)

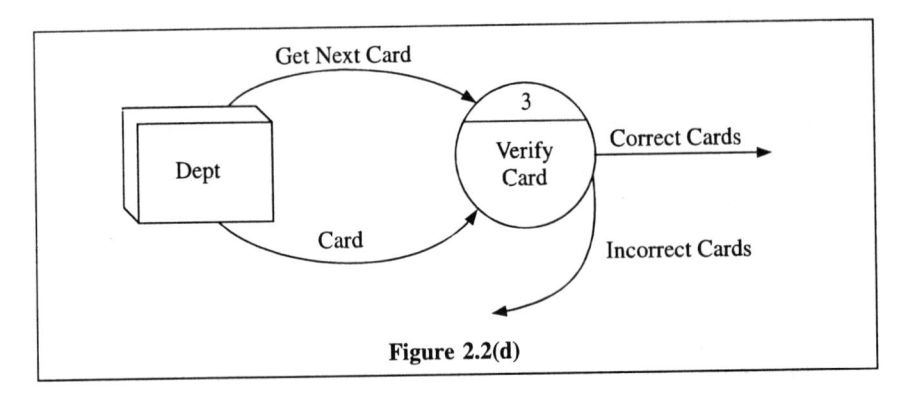

Solution to Exercise

'Get next card' is the control flow.

Levels of Data Flow Diagram

The complexity of business systems means that it is impossible to represent the operations of any system by means of single data flow diagram. At the top level, an overview of the different systems in an organization is shown by way of Context analysis diagram. When exploded into DFD, they are represented by:

- Level-0: System input/output
- Level-1: Subsystem level data flow – Functional level
- Level-2: File level detail data flow

The input and output data shown should be consistent from one level to the next.

Level-0: System input/output: A level-0 DFD describes the system-wide boundaries, detailing inputs to and outputs from the system and major processes. This diagram is similar to the combined user-level context diagram.

Level-1: Subsystem level data flow: A level-1 DFD describes the next level of detail within the system, detailing the data flows between subsystems, which make up the whole.

Level-2: File level detail data flow: A level-2 DFD details the files to which the data is applied in the system, and from which data is obtained. Each individual process is shown in detail.

Example

Figure 2.2(e) shows an organization-level context diagram for "Order Tracking System"

Application Modeling 23

[Figure 2.2(e): Context diagram showing Order Tracking System with EE1 Customer (Customer Order in, Product and Invoice out) and EE2 Supplier (Purchase Order and Payment out, Product and Invoice in)]

Figure 2.2(e)

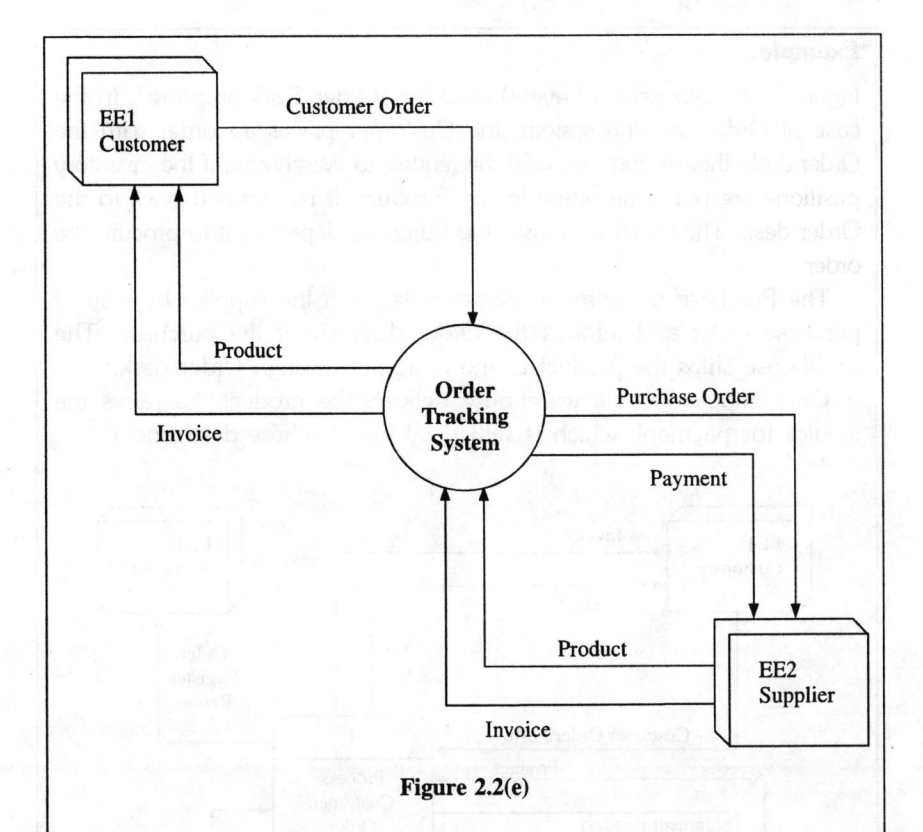

Next you expand process symbol in Figure 2.2(e). You could make out and list the basic procedures included in the order tracking system. They are:

1. Receive orders from customers
2. Send invoices to customers
3. Send products to customers
4. Send orders to suppliers
5. Receive invoices from suppliers
6. Send payments to suppliers
7. Receive products from suppliers

The above context diagram is now converted into DFD by breaking into Level-0, Level-1 and Level-2 diagrams.

Level-0 DFD diagram

The first step towards transferring the context analysis diagram to a Level-0 DFD is to list the major business process or events within the system. Link each of these processes to all the external entities, between which information flows.

24 *Workbook on Systems Analysis and Design*

Example

Figure 2.2(f) represent a Level-0 DFD for "Order Tracking Store". In the case of Order tracking system, the Customer places an order with the Order desk; they, in turn, forward the request to Warehouse. If the inventory positions are not appropriate for the product, it is informed back to the Order desk. The Order desk asks the Purchase department to procure the order.

The Purchase department, places order with the supplier by way of purchase order and informs the Order desk about the purchase. The warehouse ships the product to the customer through Order desk.

Once supplier to the warehouse delivers the product, he raises the invoice for payment, which is settled by the Purchase department.

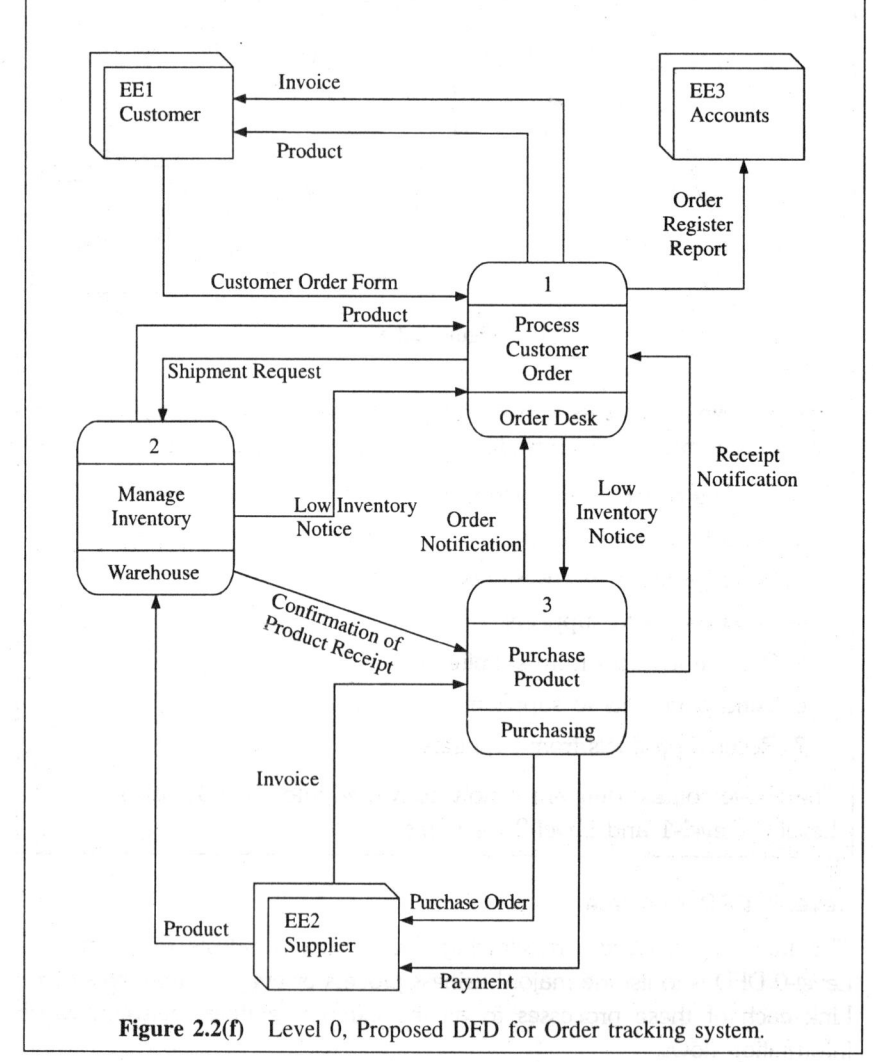

Figure 2.2(f) Level 0, Proposed DFD for Order tracking system.

Level-1 DFD diagram

In the next step, i.e. Level-1 DFD each of the main process of Level-0 is exploded into sub-process (sub-systems).

> ### Example
> Figure 2.2(g) represent the Level-1 DFD for the "Order Tracking System". In this one of the main process of the system; Warehouse is exploded. The warehouse department divides the Shipping and Receiving of the product from the supplier, and reporting of the inventory into three sub-processes.
>
> As seen in Figure 2.2(g), all other processes receiving inputs and sending outputs to the process being exploded are shown outside the boundary.
>
> **Figure 2.2(g)** Level 1 – DFD, breaking up of process "Manage Inventory".

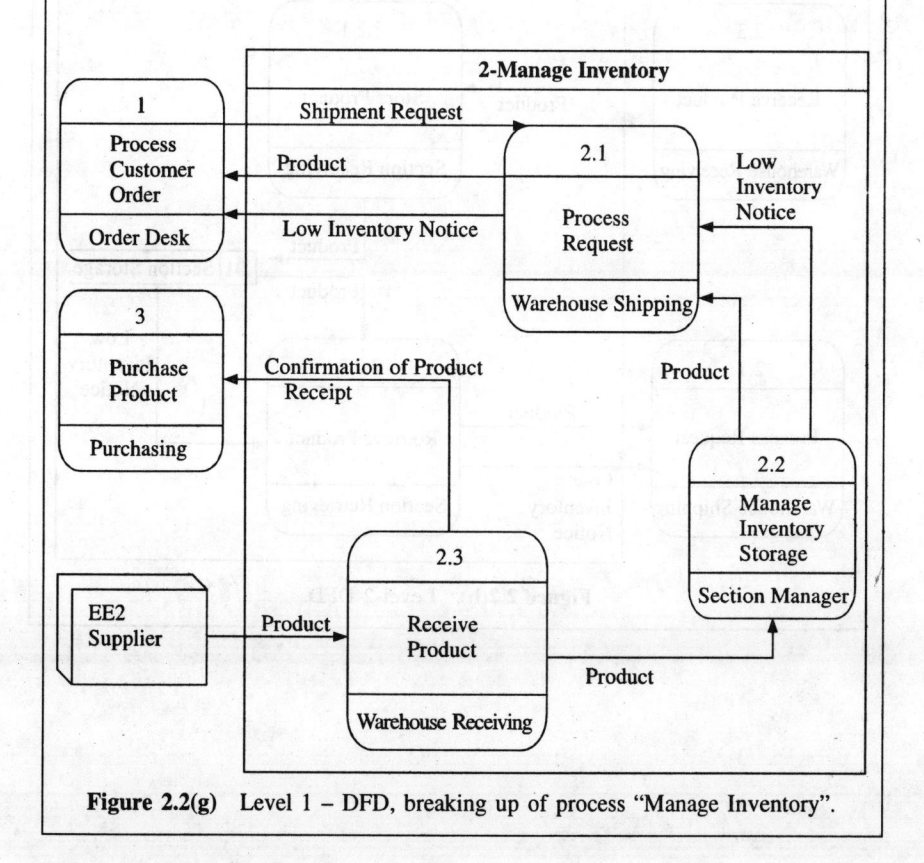

Level-2 DFD diagram

Level-2 DFD completes the process of decomposition, where each individual sub-process of Level-1 DFD is broken into details. Next identify and list the major data stores to which these processes will source their information. Link each of these processes and data stores to various external entities.

Example

As shown in Figure 2.2(h), the section receiving or retrieving the product source their information from data store D1.

As we see, each **parent process** of higher level is further broken up into **child process**, and each level explains, in precise, the information flow within the system.

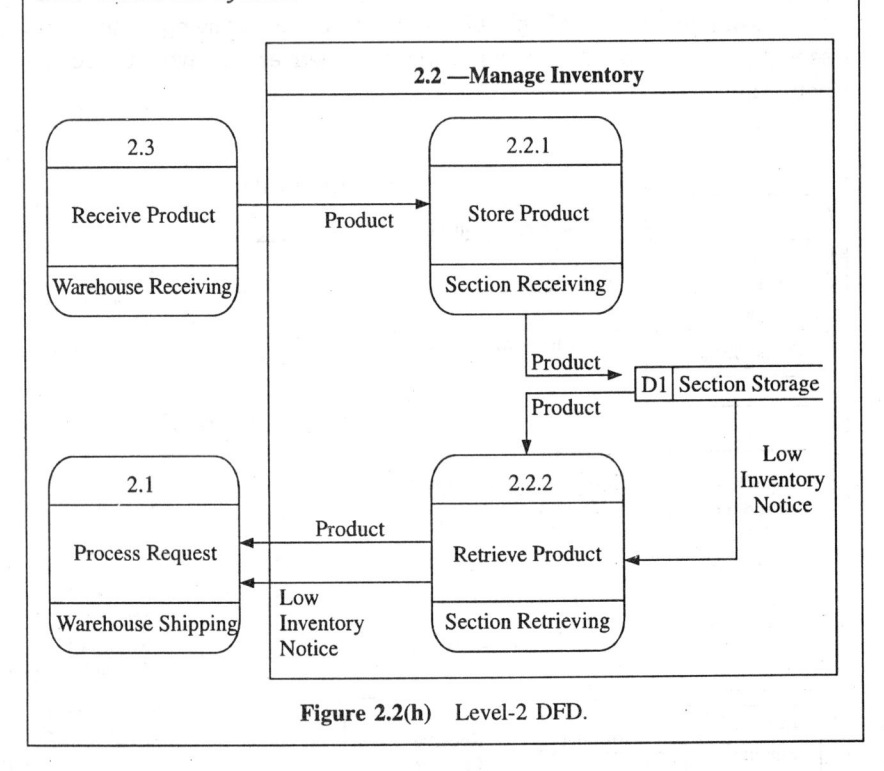

Figure 2.2(h) Level-2 DFD.

Conclusion

Models based on data flow diagrams are effective during the analysis activities of the development process. It is important to note that whenever a data flow diagram is drawn, it has to be fully exploded and a complete data dictionary is prepared. As seen, the DFD for a complicated system cannot fit into a single sheet of paper. Also packing one diagram with too many details will cause confusion.

Thus to make DFD more readable, one should decompose the context diagram into detailed DFD by following the above steps. This process is also referred as **structured decomposition.**

This way processes, data flows and data stores are represented, which form the basis for the actual development of computerized system.

Summary

A data flow diagram is a graphic tool to illustrate the manual and computer process of a computer system. It is one of the structured tools for requirement determination. Data flow diagrams focus on the data component of system and the process that transforms data.

DFD symbols are kept simple in order to convey information to the user by way of—External entities, Process, Data stores and Data flow. No reference is made to the computer hardware and other storage/retrieval devices.

A levelled set of DFD can be used to model the detailed processing that occurs in the system and to document the use of data stores. The ability of these techniques to document a system, while important, is not where their power lies. The real significance of data flow diagram is their ability to support a structured analysis process that can lead to form a model of a quality system.

Exercise

Figure 2.2(i), illustrates a context diagram for student registration system. From this diagram it is clear, for example, that a student submits a registration request to the system and may receive a class schedule, a tuition bill, and an accounts receivable statement from the system. Thus the context diagram knows the scope of the whole system.

Explode the context analysis diagram in Figure 2.2(i), into Level-0, Level-1 and Level-2 DFD.

Solution to Exercise
Level-0

As shown in Figure 2.2(j), the student registration process consists of three sub-systems, namely STUDENT REGISTRATION, SCHEDULE PREPARATION and ACCOUNTS RECEIVABLE.

28 Workbook on Systems Analysis and Design

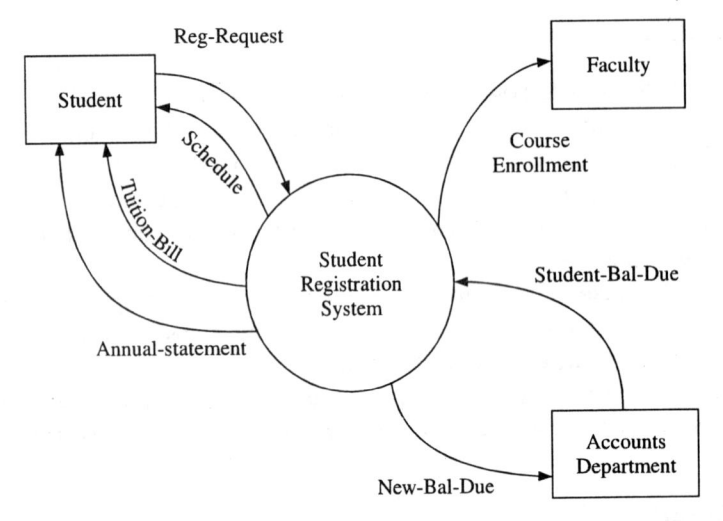

Figure 2.2(i) Context diagram student registration system.

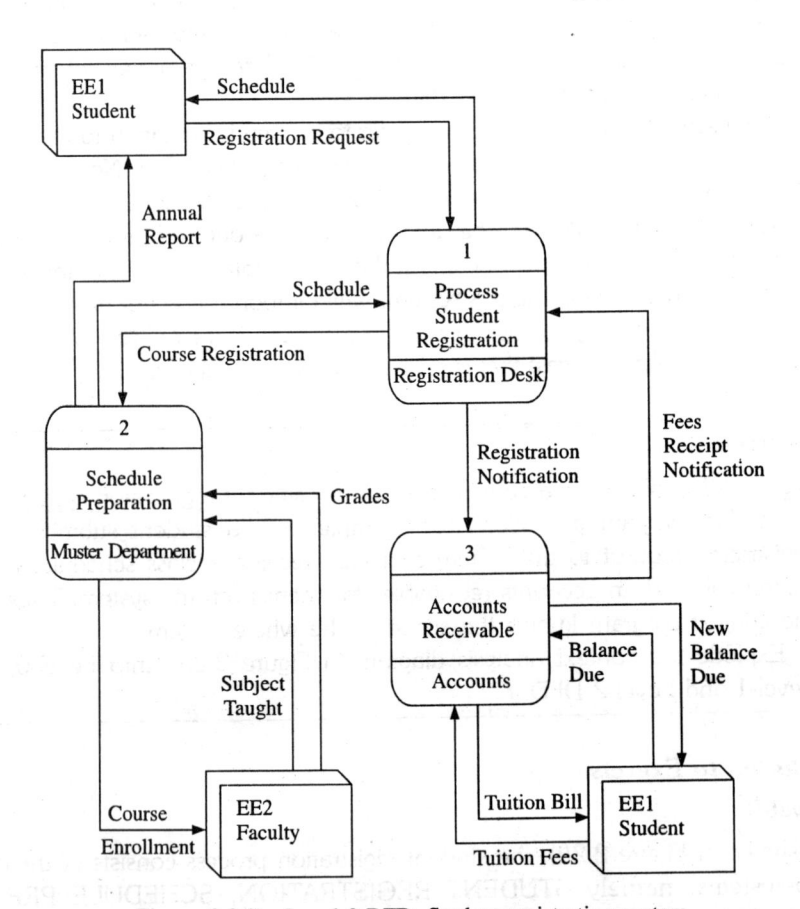

Figure 2.2(j) Level-0 DFD: Student registration system.

The student registration is handled by registration desk. The process handles course enquiry from the student and registers the student for the course.

Muster department does the Schedule preparation. The process allocates faculty to various courses and works out the schedule for each course.

Accounts department handles the Accounts Receivable sub-system. It collects the tuition fees from the student based on the course registered.

Level-1

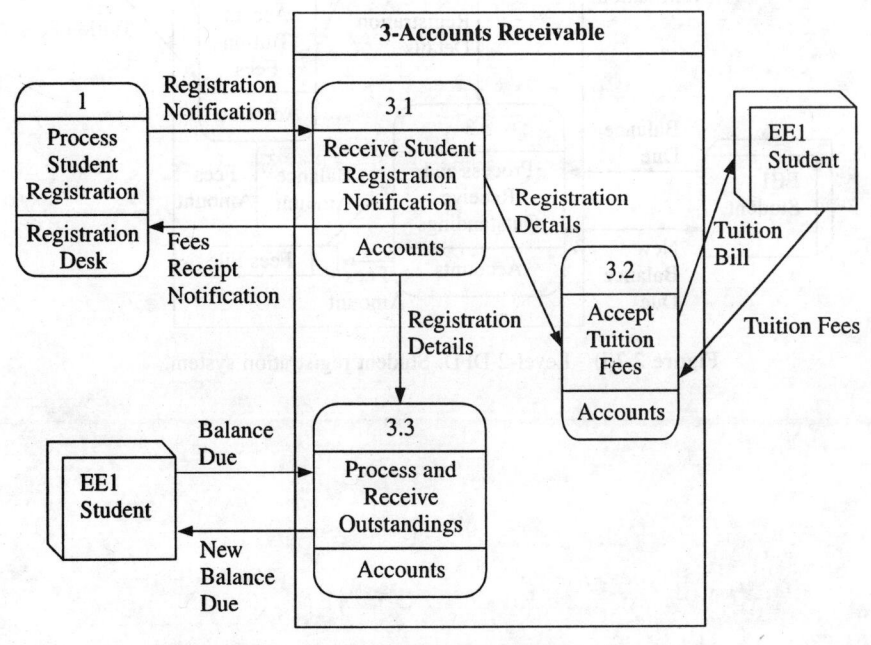

Figure 2.2(k) Level-1 DFD: Student registration system.

As shown in Figure 2.2(k), one of the modules, Accounts Receivable is exploded. There are further three sub-processes in this module, Receive Student Registration Notification, Accept Tuition Fees, Process and Receive Fees Outstanding.

Registration desk is the interface to Accounts Receivable process.

Level-2

Level-2 DFD completes the process of decomposition, where for each sub processes the major data stores to which these process will source their information are identified and linked. Link each of these processes and data stores to various external entities.

As shown in Figure 2.2(l), all the processes of Accounts Receivable receive and update data to Students file and Fees file. On similar lines you will explode the other two main process as shown in Figure 2.2(j), viz. STUDENT REGISTRATION and SCHEDULE and PREPARATION

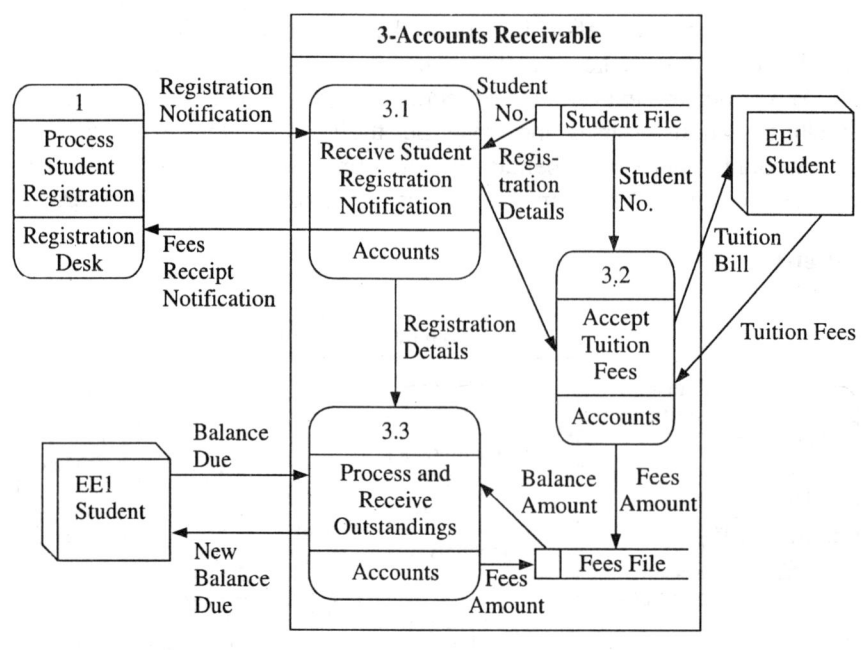

Figure 2.2(l) Level-2 DFD: Student registration system.

Application Modeling   31

Practice for Section 2.2

#1
Draw a DFD to include the cheque verification process based on **context analysis diagram** in Figure 2.2(a).

#2
Draw a context analysis diagram and a data flow diagram for the front office system in a 5-star hotel described below:

Front office of any hotel is responsible for all room reservations, room allocations and final settlement of bills. Any company or person can reserve rooms for their future stay. They have to indicate from what date to what date they need the room. They also have to indicate how many rooms are required. Sometimes the reservations could be cancelled or the dates or number of rooms changed. For reservation, cancellation or modification of rooms, customer receives an acknowledgement from the hotel.

#3
Draw the context diagram and the data flow diagram for the following maintenance system:

Zen Systems is a computer hardware manufacturer selling complete range of PCs. This company also maintains the systems they sell after entering into annual maintenance contract. Every year the contract is to be renewed for a specific amount.

To attend to customer complaints the company has on its rolls a large number of hardware engineers. As and when a complaint is received by the maintenance system, they are entered in a form after verifying for valid customer and the machine. These forms are collated and a list is prepared which is handed over to hardware engineers in the morning.

During the maintenance of these systems, hardware engineers use spare parts, which they get issued from the store. After hardware engineers attend to customer calls, complaint and parts used are passed on to the maintenance system where they are updated. Stock reports are sent to stores every month by the system.

#4
In Cybertech Systems, after the Finance department receives attendance details from all other departments, they calculate the total salary to be paid to the employees. Total salary is calculated based on how many days each employee has worked.

Once the total salary figures are ready, the Finance department calculates the tax to be deducted. The salary sheet is sent to the bank for the employee's account to be credited. Pay slips are then printed and given to departments for handing them over to the employees.

Based on the above description:

(a) **Identify the processes with respect to Finance department.**
(b) **Identify the external entities for the processing of Cybertech Systems & Software salary system.**
(c) **With respect to other departments, identify data flows to and from the Finance department.**
(d) **Draw a high-level context diagram for the above system.**
(e) **Draw data flow diagram to represent the finance department processes.**

#5
Draw a data flow diagram to represent a simplified inventory system whose main processing is described:

When parts are received from VENDORS, an INVOICE accompanies them. The invoice is first checked against the ORDER file to verify that the parts were actually ordered. If they were ordered, the PART-QUANTITY for those parts are updated in the INVENTORY file.

For accepted parts, PAYMENT is sent to vendors, and the payment transactions are entered in the GENERAL LEDGER file. To check parts out of inventory, people on the shop floor submit a REQUISITION. This requisition form is used to update inventory file. Each week, the complete inventory is processed to identify parts whose part-quantity has fallen below the REORDER-LEVEL. For each part, a PURCHASE-REQUEST is sent to the PURCHASE department.

#6
Draw the context analysis diagram and top level DFD based on the following information for a manufacturing system:

Customer sends enquiry to Commercial department, receives quotation from the Sales department and places an order. Based on the customer order, the work order is sent to the Planning department for planning scheduling and control; in turn, the Planning department raises a Job order on the 'Shop floor'. On completion, delivery note and invoice are made out. Costing department also prepares an order-wise comparative statement of estimated and actual costs.

#7
Century Systems is the statistics provider organization of all the first class cricket matches conducted across the world round the year. They compile all the statistics of the players in terms of the matches played, venues, country, match type: if test or one-day or three-day. For each match statistics is compiled for each player, runs scored in each innings, wickets and catches taken, run out effected, number of fours and sixes hit, etc.

Application Modeling 33

The company wants to develop a computer application to compile all these statistics. Work out a logical design using context diagram and a DFD.

#8
List all the errors you observe in the data flow diagram illustrated in Figure 2.2(m).

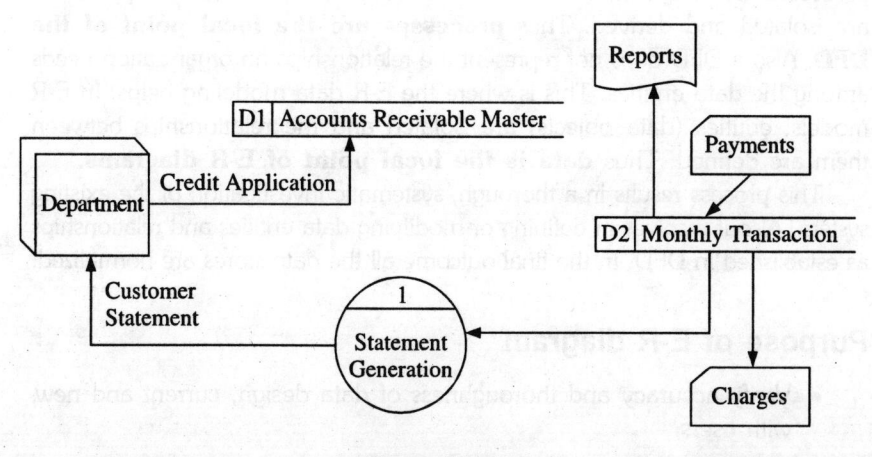

Figure 2.2(m)

2.3 E-R DIAGRAMS

Entity-relationship (E-R) diagrams are graphic illustrations used to display objects or events within a system and their relationships to one another. E-R diagrams model data in much the same way as DFDs model processes and data flows.

Why is a data flow diagram not a complete data description? In DFD, processes are first identified and then the data flow between the processes are isolated and derived. Thus **processes are the focal point of the DFD**. Also a DFD does not represent the relationships an organization needs among the data entities. This is where the E-R data modeling helps. In E-R models, entities (data objects) are isolated and the relationships between them are defined. Thus **data is the focal point of E-R diagrams**.

This process results in a thorough, systematic investigation of the existing system. Also it resolves in defining or modifying data entities and relationships as established in DFD. In the final outcome all the data stores are normalized.

Purpose of E-R diagram

- Verify accuracy and thoroughness of data design, current and new, with users.
- Organize and record organizational data entities, relationships and scope through decomposition and layering.
- Enhance the overall communication between development project team members, system technicians, management and users with the use of graphic models.
- Generally simplify and bolster the creative data design process.

How to represent E-R diagram

Figure 2.3(a) illustrates an E-R diagram. An E-R model represents entities and the relationships that exist between them. A box symbol as shown in the figure is used to represent Entities and a diamond for Relationship.

ENTITY ←— Relationship —→ ENTITY

Figure 2.3(a) An E-R model.

Generally speaking, the more "event-like" a relationship is, more likely it will be to require a unique data structure or data table. An important part of definition of a relationship is its cardinality. Cardinality specifies how many instances of one entity can describe one instance of the other entity in the relationship. This will be our next topic of discussion.

Types of E-R Relationships

There are **three basic types** of relationships modeled between entities on an E-R diagram. They are:

- **One-to-one**
- **One-to-many**
- **Many-to-many**

> *Example*
>
> Consider a student registration system, where the entities identified in data flow diagram are STUDENTS, INSTRUCTOR, COURSES OFFERED, and COURSE SCHEDULE.

We see how relationships can be established between these entities.

One-to-one

A **one-to-one** relationship signifies that items in one of the associated entities have exactly a one-to-one correspondence to items in the other entity.

Many a time, a one-to-one relationship can be resolved by merging the entities. For instance, in the example here if the instructor were always assigned to same course every year, it might be useful to merely merge the entities.

```
INSTRUCTOR ◄── ⟨ Assigned to ⟩ ──► COURSES OFFERED
```

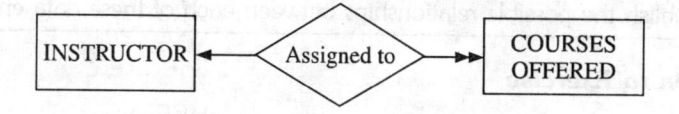

Figure 2.3(b) A one-to-one relationship.

One-to-many

A **one-to-many** relationship implies that more than one item in one entity may be associated with many items in another entity. One item in an entity refers to many items in another, but not vice versa.

In the above example, as soon as an instructor teaches more than one course in any given year, this becomes one-to-many relationship. As a rule, never merge any entities where it seems possible for a one-to-one relationship to become a one-to-many.

```
INSTRUCTOR ◄── ⟨ Assigned to ⟩ ──► COURSES OFFERED
```

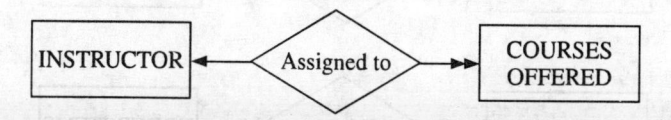

Figure 2.3(c) A one-to-many relationship.

Many-to-many

A **many-to-many** relationship implies that each item in an entity may be associated with many items in another entity and vice versa.

As shown in the Figure 2.3(d), students register for many courses, which represents the case of many-to-many relationship. Also one course will be taken up by many students.

```
STUDENTS ◄──< Register for >──► COURSES OFFERED
```

Figure 2.3(d) A many-to-many relationship.

Exercise

Consider the Order Tracking system.

Customer place orders for universal products. Orders are filled in the Order Processing department by order processing clerks. In the Order Processing department, an Order number is assigned to each order for identification and an invoice with the cost of the products for the order is produced. When the invoice is sent to the customer, a shipment is also made to the customer by the Shipping department. After the DFD are drawn, the following are the data entities established: SHIPMENT, CUSTOMER, ORDER, INVOICE and PRODUCT.

Establish the possible relationships between each of these data entities.

Solution to Exercise

The following diagram represents a one-to-one relationship.

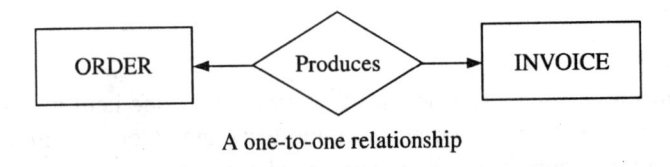

```
ORDER ◄──< Produces >──► INVOICE
```

A one-to-one relationship

The relationships shown here are all one-to-many relationship.

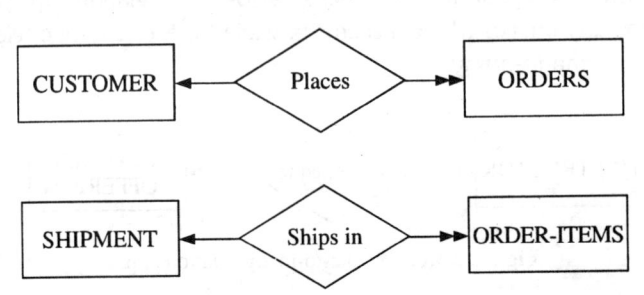

```
CUSTOMER ◄──< Places >──► ORDERS
```

```
SHIPMENT ◄──< Ships in >──► ORDER-ITEMS
```

```
PRODUCT  ◄── Ordered ──►► ORDER-ITEM

ORDER    ◄── Contains ──►► ORDER-ITEMS

CUSTOMER ◄── Receives ──►► SHIPMENTS

PRODUCT  ◄── Ordered ──►► ORDER-ITEM
```

A one-to-many relationships

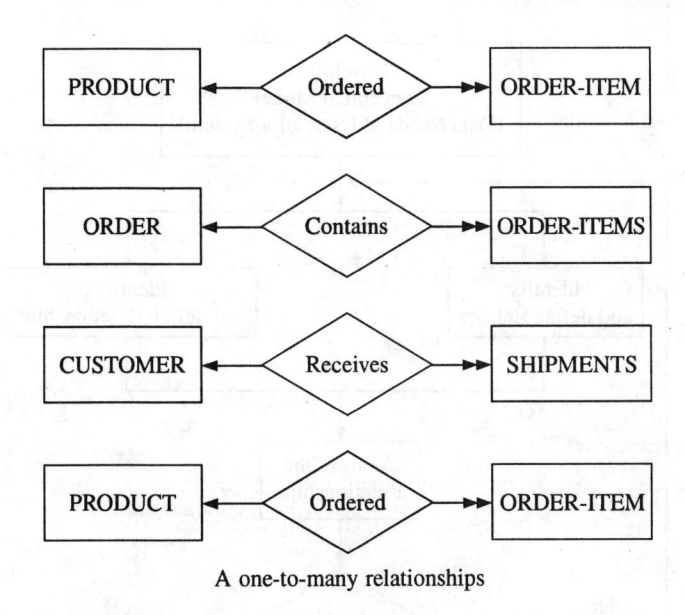

Steps in Building an E-R Diagram

There are 6 steps to follow in building an E-R diagram.

- Determine the data entities.
- Generate a list of potential entity relationships or pairings.
- Determine the relationship between the entity and pairings.
- Analyze the significant entity relationships.
- Develop an integrated E-R diagram.
- Define and group the attributes for each data entity.

This top-down approach is applied to the construction of a data model by way of E-R diagram.

The above steps are illustrated as a flowchart in Figure 2.3(e).

Determine the data entities

The data entities, are those objects within the system, which are the likely candidates to have information about them stored. Thus from the DFD all the data stores which you have identified becomes the choice for data entities.

Generate a list of potential entity relationships or pairings

What is the maximum number of potential combinations among different individual entities. Each such combination creates a pairings among the entities.

Determine the relationship between the entity and pairings

The relationship of each should be examined both for logical association's

38 Workbook on Systems Analysis and Design

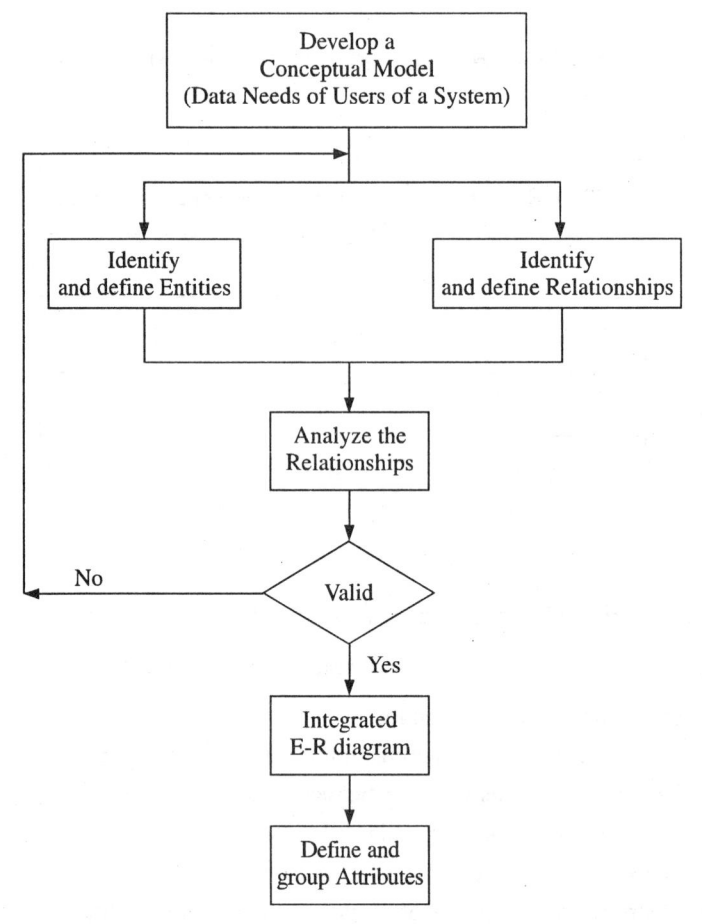

Figure 2.3(e) Procedure for data modeling through E-R diagram.

nature and for significance. If an association is determined to be insignificant, i.e. it requires no data integration or linkage, then it can be removed from the data model entirely.

Example

Let us take the example of Student registration system, where the data entities are STUDENTS, INSTRUCTORS, COURSES, SCHEDULES, and look at all the possible pairings:

Students to Instructors: Students taught by instructors.

Linkage required for knowing to which instructors a student is assigned

Students to Student schedules: Students generate schedule upon registration.

Linkage required for knowing to which courses a student has scheduled for a given term.

Students to Courses offered: Students select the courses offered.
Linkage required for knowing details about courses for which a student will register.

Instructors to Student schedules: Instructor listed on student schedule.
This relationship is not significant, as the instructor is not listed on the course schedule.

Instructors to Courses offered: Instructor teaches courses.
This relationship establishes link to know which instructor teaches which courses.

Student schedules to Courses offered: Student schedules list selected course offerings.
Linkages to know for which courses a student has registered.

Analyze the significant entity relationships

A single E-R diagram should be developed for each significant pairing, with the type of relationship (one-to-one, one-to-many, many-to-many).

Example

Let us take the example of student registration system and looking at the pairings, establish the relationships between entities.

Students to Instructors: Students taught by instructors.
Relationship is Many-to-many.

Students to Student schedules: Students generate schedule upon registration.
Relationship is Many-to-many.

Students to Courses offered: Students select the courses offered.
Relationship is Many-to-many.

Instructors to Student schedules: Instructor listed on student schedule.
This relationship is not significant.

Instructors to Courses offered: Instructor teaches courses.
Relationship is One-to-many.

Student schedules to Courses offered: Student schedules list selected course offerings.
Relationship is Many-to-many.

Develop an Integrated E-R Diagram

Finally all the individual E-R diagram/pairings are assembled to represent a single diagram as shown in Figure 2.3(f)

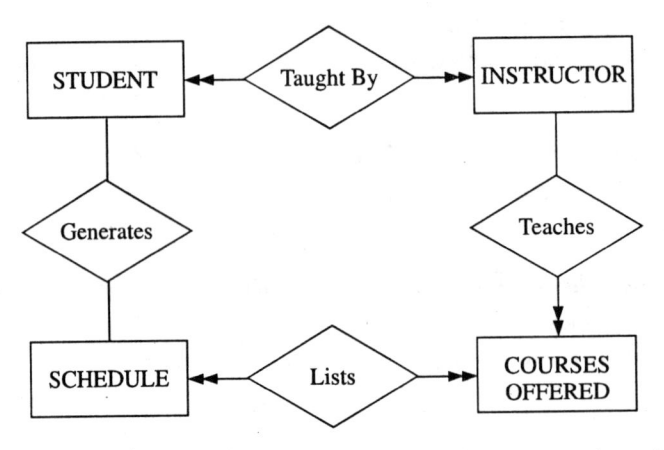

Figure 2.3(f) Final E-R diagram representing relationship among students, instructors and courses offered.

Next we define and group the attributes for each data entities, as shown below:

COURSES OFFERED = Class-Number
　　　　　　　　　　Class-Name
　　　　　　　　　　Class-Credits
　　　　　　　　　　Class-Room
　　　　　　　　　　Class-Time
　　　　　　　　　　Class-Instructor
　　　　　　　　　　Class-Enrollment
　　　　　　　　　　Class-Maximum-Limit

INSTRUCTOR = Instructor-Number
　　　　　　　Instructor-Name
　　　　　　　Instructor-Department
　　　　　　　Instructor-schedule (for all classes taught)
　　　　　　　　　{Class-Number}
　　　　　　　　　{Class-Name}
　　　　　　　　　{Class-Credits}
　　　　　　　　　{Class-Enrollment}

STUDENT = Student-Number
　　　　　　Student-Name
　　　　　　Student-Address
　　　　　　Student-Level

```
                    Student-Credits-Earned (for all classes)
                        {Class-Number}
                        {Class-Name}
                        {Class-Credits}
                        {Class-Grade}
REGISTRATION =      Student-Number
                    Class-Number

SCHEDULE =          Instructor-Number
                    Class-Number

ATTENDANCE =        Class-Number
                    Class-Name
                    Class-Credits
                    Class-Room
                    Class-Time
                    Class-Instructor
                    Class-Attendance (All students present/absent)
                        {Student-Number}
                        {Student-Name}
                        {Student-Level}
```

The above is the partial data dictionary list arrived for Student Registration system. This will be then converted into normalized data-structures.

Summary

Conceptual data modeling consists of analyzing and describing the data needed by the users of the system. E-R diagram establishes the data model by way of relationships between two data entities. The constructs of E-R diagram are entity, identifier, attribute, and relationship.

The E-R diagram starts with construction of high-level model that is partitioned into subsystems. After this, entities and relationships are established. This relationship can be one-to-one, one-to-many and many-to-many. After all subsystem data models have been constructed into individual E-R diagrams, they are then integrated into one integrated E-R diagram for the entire system.

Practice for Section 2.3

#1
Hale Heart hospital has a pharmacy division, which procures medicines/drugs from various pharmacists. The division also maintains the drug's inventory and issues medicines to patients as prescribed by the different Doctors in the hospital. The patients have to pay for the medicines if they do not come under the welfare category.

After the system study was carried out, the analyst identified five major entities and their relationships. They were: DRUGS, PHARMACIST, PRESCRIPTONS, PATIENTS, PHYSICIANS,

Establish the relationships between the data entities, and draw a E-R diagram.

#2
In the pharmacy data model of #1, the cardinality of the relationship between drug and supplier supplying the drug is many to many. Suppose Hale Heart hospital designates one supplier as the primary supplier for a drug, but also uses other suppliers when necessary.

Change the model of #1 to reflect the above relationship.

#3
Century Systems is the statistics provider organization, keeping track of the all first class cricket tournament conducted across the world round the year. They compile all the statistics of the players in terms of the matches played, venues, country, match type: whether test or one-day or three-day match. For each match, statistics is compiled for each player, runs scored in each innings, wickets and catches taken, run out effected, no. of fours and sixes hit, etc.

After the system study was carried out, the analyst identified seven major entities and their relationships. They were: PLAYER, CENTURY, GROUND, MEDIA, COMPANY, ADVERTISEMENT and CAMPAIGN.

Establish the relationships between the data entities, and draw a E-R diagram.

#4
What is an entity-relationship (E-R) diagram? What is the purpose for developing this type of diagram?

#5
What are the three relationship types of E-R diagram? How are these relationships paired to build an E-R diagram?

3

Database Design

In the second chapter, the conceptual data model was introduced resulting in constructs of E-R diagram. The E-R diagram describes the data requirements of an information system, however is not a complete database description. A complete database description must conform to the rules of logical and physical data structures imposed by the database management system that will manage the database. Thus the next step is to look into physical database design, transform the logical description into a physical model that describes how database will be organized and accessed on physical storage devices.

3.1 DATA DICTIONARY

One of the key requirements in the development of DFD is the precise naming of the components. This is especially critical for the data oriented components, namely the **data flows** and **data stores.** Therefore it is necessary to establish **data dictionary**, that contains the names assigned to all data flows and data stores, with precise and complete definitions for each term.

In a data dictionary, data flows and data stores are described in terms of data elements and data structures. A **data element** is a basic unit of data that can be assigned a meaningful data. **Data structures** are made up of data elements and other sub data structures or a mixture of both.

Example

Consider a data store maintained by a cargo service company. This data store would contain details regarding parcel type and weight, sender details, destination information and freight charges.

Given below is the data store broken down into its component parts.

PARCEL-DETAILS
    Parcel-No.
    Parcel-Type
    Parcel-Weight

> SENDER-DETAILS
>   From-Name
>   From-Address
>   From-City
>   From-Pin
> DESTINATION-DETAILS
>   To-Name
>   To-Address
>   To-City
>   To-Pin
> Freight-Amount
>
> Here PARCEL-DETAILS is the data structure. This data structure is made up of four data elements (Parcel-No., Parcel-Type, Parcel-Weight and Freight-Amount).
> But SENDER-DETAILS and DESTINATION-DETAILS are two other data structures within PARCEL-DETAILS.
> SENDER-DETAILS is further made up of four data elements From-Name, From-Address, From-City, From-Pin respectively.
> Similarly data structures DESTINATION-DETAILS is made up of To-Name, To-Address, To-City, To-Pin data elements.

Data Flows and Data Stores

As stated above, data flows and data stores are made up of data structures. Data flows are paths or 'pipelines', along which data structures travel, whereas data stores are places where data structures are kept until needed.

Data flows are data structures in motion, while data stores are data structures at rest.

Hence it is possible that a data flow and a data store would be made up of the same data structure.

Representing Data Structures

Its very important to understand the notation used for writing data structures. Consider the structure of an INVOICE.

INVOICE
   Invoice-No
   {Order-No}
   {Challan-No}
   Invoice Date
   Customer-No
   Customer-Name

Item-Details * (n-m)
    Item No
    Item Description
    Qty-Sold
    Rate
    [Discount]
    Total Value
    Invoice Amount

There are some symbols used in the above structure:

* (n-m) The * represents a repeating group (Item-Details), with all the elements (Item-No, Item Description till Total Value).

    (n-m) means there be iterations (number of occurrences) from *n* to *m* times, with *n* being at least 1.

    If the upper limit of the occurrence cannot be specified then it is left blank.

[ ]   Represents an optional element (Discount) which may or may not take values.

{ }   Indicates that any one value will be stated for each record. In this case, either the Order-No or the Challan-No will be stated for each invoice.

Examples

Assuming there can be a maximum of 6 subjects a student can take in a year. This will be represented as: Course-name*(1-5).

If the number of items, which can be ordered by a single order is unlimited, it is represented as: Items*(1-).

If the data element, 'Person-Signed' in Purchase Order need not be present, then it will be represented as: [Person-Signed].

Exercises

Answer if the following are true or false, with reasons.

#1
Quantity *(1-10) means that the group 'Quantity' could be repeated anywhere between 3 to 10 times in a data structure.

#2
Name *(0-4) would mean that there may be no occurrence of Name in the data structure.

#3
[Cheque No] indicates that the Cheque No must exist in the data structure containing the element.

#4
{Present}
{Absent} indicates that one or more of these elements in a document can be represented.

#5
Represent the report shown below as a data structure.

ATTENDANCE REGISTER FOR 18th May 1998

Employee code	Employee name	Salary	Conveyance	Present days	Absent days
0123	Reshmi Nair	8,567		30	0
0144	Srinivasan S.	15,654	5000	25	5
1234	Jiju John	10,341		28	2
2356	Lata Chablani	18,946	3000	29	1

Solution to Exercises

#1 True
   Because (3-10) shows occurrence between 1 to 10 times.

#2 True
   Because (0-) would indicate no occurrence of an element.

#3 False
   Square brackets [ ] indicates that the element may or may not have values.

#4 True
   As per the property of { }.

#5 ATTENDENCE REGISTER
Date
Employee Details *(1-)
   Employee Code
   Employee Name
   Salary
   [Conveyance]
   {Present}
   {Absent}

Database Design 47

Practice for Section 3.1

#1
Pick, the data elements and data stores from the following:
- (a) Attendance-Details
- (b) Employee-Name
- (c) Employee-No
- (d) Days-Present
- (e) Invoice
- (f) All of the above

#2
Identify data elements and data structures of the PURCHASE ORDER data structure as illustrated below.

```
PURCHASE-ORDER
    Date-of-order
    Order-No
    CUSTOMER-DETAILS
        Company Name
        Phone-No
        SHIP-TO-ADDRESS
            Name
            Street
            Area
            City
            Pin
    ITEM-DETAILS
        Item-Number
        Quantity
    Order-prepared-by
```

#3
In the above structure list the data elements and data structures in CUSTOMER-DETAILS.

#4
Data dictionary contains an entry of the following:
- (a) External Entity
- (b) Data flow
- (c) Data store
- (d) Process
- (e) All of the above
- (f) None of the above

Choose the correct option

#5

Consider the following portion of a DFD describing the portion of the reservation process at Hotel desk, shown in Figure 3.1. Would there be any difference between the data structures of the data flow "reservation" and the data store "reservation".

Hint: Data flow, flowing through pipeline and data structure at rest.

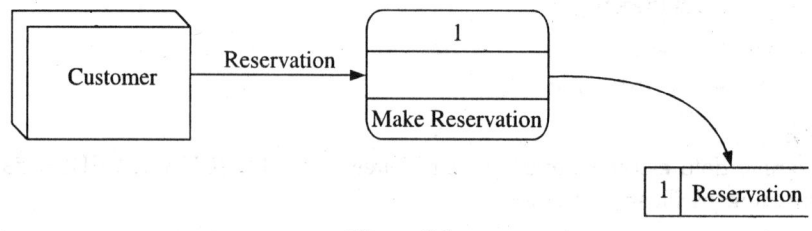

Figure 3.1

#6

Modify the following data structure,

PAYMENT ADVICE
    Date
    Name of supplier
    Address of supplier
    Cheque number
    Bank payment details
        Bank name of supplier
        Bank account number of supplier
    Invoice*(1-)
        Number
        Date
        Amount
        [Description]
    Payment total

such that

1. "Supplier code" can be represented instead of "name of supplier" and "address of supplier".
2. "Supplier code" has an optional "credit status" data element. "Credit status" can have an optional set of one to five "number of days payment delayed".
3. "Address of supplier" has one or two optional phone numbers, fax numbers.
4. "Invoice" is itself an optional data structure.

3.2 LOGICAL DATA ANALYSIS

Once data stores and data structures have been isolated and defined, the next step is to examine and refine the actual organization of the data stores. Logical Data Analysis provides a means for organizing logically simple and efficient data stores.

Two main criteria are applied to data stores to determine whether their components and organization are optimum for the system they support.

Simplicity. The goal is to structure a data store so that it can be implemented as a simple sequential or direct-access file. This means, only a primary key should reference the components of the data store, and there should be no repeating groups of data within the data store.

Redundancy. The second criterion for logical data design is non-redundancy. Redundancy occurs, when the same data element has been placed within two or more data stores. Redundancy can threaten the integrity of a system. The task at hand, therefore, is to derive a logical structure for the proposed system. The procedure used to derive this logical structure is called **normalization.** The advantages of normalization are ease of understanding, ease of use, ease of implementation, and ease of maintenance.

The most important benefit of designing a database through process of normalization from simple entities put together, is that the database will be easy to change as new modules are further developed into applications, thereby making it a bottom-up approach, that can handle incremental growth.

Designing the Database

The logical and physical database structures guide the design of a database. The steps in designing a database for a system are as follows:

1. Designing the logical database.
   - Transform the conceptual data model into a logical database design.
   - Evaluate and change the logical design for flexibility.
   - Determine volumes of record instances.
   - Evaluate and change the logical design for efficiency.
   - Specify the logical design using DDL (Data definition language) of DBMS concept.
2. Designing the physical database.
   - Evaluate and change physical design for efficiency.
   - Evaluate the database for integrity.
   - Add integrity controls to the database design.

When the design is complete, the deliverables will include the DDL for the logical and physical data structures of the database.

Functional Dependency

To understand normalization, one must understand the concept of functional dependency. A data element is functionally dependent on the key of a record if the value of the key determines a unique value for the data element.

Suppose we have two data elements A and B, then each value of data element is always associated with one single value of data element B. We can also say A is a determinant of element B, B is determined by A or B is functionally dependent on A. To say A determines B means that if we are given a value of A we can always find, unambiguously, the value of B.

How to establish function dependency

Example

In this example we will see how five data elements related to customer's order are functionally dependent.

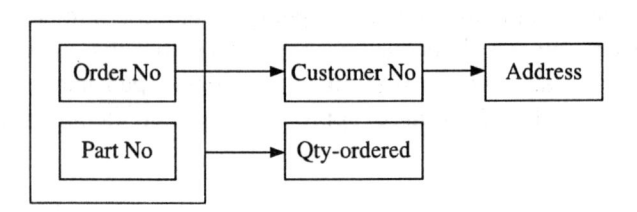

Order No → Customer No → Address

Part No → Qty-ordered

Figure 3.2(a)

(a) As seen in Figure 3.2(a), each **Order No** is associated with only one **Customer No**, but a Customer No may be associated with many Order No.

(b) Each Customer No is associated with only one **Address**, but the same Address may be associated with more than one Customer No.

(c) Each Order No may be associated with many **Part Nos**, and each Part No can be associated with many Order Nos.

(d) Each Order No + Part No taken together is associated with only one Quantity ordered, but a **Qty-ordered** may be associated with more than one Order No + Part No.

From these enterprise rules, we can say Order No is a determinant of Customer No, Customer No is determinant of Address, and Order No + Part No is determinant of Qty-ordered.

Database Design 51

From the above explanation—A determines B if a given value of A is always associated with the same single value of B—can also be interpreted in functional dependency as:

"**Given a relation R, attribute A of R is functionally dependent on attribute B of R if and only if each B-value in R has associated with it precisely one A-value in R (at any one time). The attributes A and B may be composite (keys) also**".

In the above example an Order No is associated with a set of Part No. Thus a multi-valued relationship is a particular kind of relationship in which instead of one element determining a single value of another element, it determines a set of values.

$$A \longrightarrow B$$

implies B is functionally dependent on A

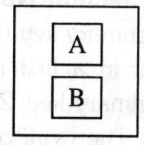

$$\begin{array}{c} A \\ B \end{array}$$

implies A and B have multilevel dependencies.

Scope of dependency

Functional dependency is about the meaning of data. It is the rules which express the meaning of the data and which must be translated into a relational model. Finally, functional dependencies express **constraints**. To say that Order No determines Customer No, expresses the constraint that in the world of the enterprise in question, an order (represented by the data element Order No) is always associated with one customer (represented by the data element Customer No).

This way functional dependency is used to describe the relationship between data elements and the key of a record.

Process of Normalization

In the DFD the system showed several data stores, some used to maintain data over a period of time and others used as transitional stores for the production of reports. During modeling process, each data store was designed to support a particular transformation. Each of these data stores is an iteration of a data structure. Thus, the normalization of data stores is essentially equivalent to the normalization of a set of data structures.

Data stores are said to be un-normalized when they contain repeating groups. There are **three** steps/stages in this process:

- First Normal Form
- Second Normal Form
- Third Normal Form

Some methodologies may carry the level of normalization further.

First normal form

A data structure is said to be in **First Normal Form** when it contains no repeating groups. For this, partition each data structure containing repeating groups that accomplish the same purpose.

Second normal form

A data structure is said to be in **Second Normal Form** when all its non-key fields are dependent on the primary key (composite key). For this, verify that each non-key data element in a first normal form structure is fully functionally dependent on the primary key. 2NF apply only to records that have compound/composite keys. The work of this step is accomplished by verifying that each non-key field in a data structure is dependent on the full concatenated key, and not just on a partial key.

In case a data element is determined uniquely by only a part of the key, the element should be removed from the structure and placed in a structure of its own.

Third normal form

A data structure is said to be in **Third Normal Form** when all its non-key fields are independent of one another. Verify that all non-key data elements in 2NF data structure are mutually independent of one another.

In other words, there is a check for redundancy within the relation. Duplicate data elements or elements that can de derived from other elements are removed at this stage. After the set of data structures has been put in 3NF, there are likely to be redundancies among the normalized structures, like some data elements repeating in different data stores. This must also be normalized at this stage.

Example 1

We will see how the ORDER-REPORT shown in Figure 3.2(b) will be fully normalized into first, second, and third normal form.

ORDER-REPORT

Order No	Order Date	Invoice No	Invoice Date	Product No	Product Description	Quantity Ordered
25	15/8/97	1568	28/8/97	4567	Computer forms	5
128	16/8/97	1569	28/8/97	4568	Printer ribbons	10
535	20/8/97	1570	28/8/97	4568	Printer ribbons	10
				4567	Computer forms	20
				4555	A4 stationery	45
				4550	Size1 pads	25

Figure 3.2(b) Consolidated order report.

The above order report is not normalized. The first step is to convert this into a normalized form. For this, fill in all the repeating elements, namely Order No, Order Date, Invoice No, Invoice Date. This translates into Figure 3.2(c).

In the Figure 3.2(c), a value for the key Order No will always provide a unique value for Order Date, because an order is placed on only one particular date. Thus we say that Order Date is functionally dependent on Order No. In contrast, many values of Product Description are possible for a value of Order No as seen in Figure 3.2(c).

We will now convert the order-report in Figure 3.2(c) ino 1NF.

First Normal Form (1NF)

A record is in first normal form (1NF) if there are no data elements that repeat a variable number of times. Thus in Figure 3.2(c), for a given order, Order No, Order Date, Invoice No and Invoice Date repeats variable number of times.

Order No	Order Date	Invoice No	Invoice Date	Product No	Product Description	Quantity Ordered
25	15/8/97	1568	28/8/97	4567	Computer forms	5
128	16/8/97	1569	28/8/97	4568	Printer ribbons	10
535	20/8/97	1570	28/8/97	4568	Printer ribbons	10
535	20/8/97	1570	28/8/97	4567	Computer forms	20
535	20/8/97	1570	28/8/97	4555	A4 stationery	45
535	20/8/97	1570	28/8/97	4550	Size1 pads	25

Figure 3.2(c) Normalized order report.

Further, when looking into product details, Product Description is not functionally dependent on Order No; it is, however functionally dependent on Product No. This way the dependencies are established and the process of normalization starts.

ORDER RECORD

Order No	Order Date	Invoice No	Invoice Date
25	15/8/97	1568	28/8/97
128	16/8/97	1569	28/8/97
535	20/8/97	1570	28/8/97

ORDER-ITEM

Order No	Product No	Product Description	Quantity Ordered
25	4567	Computer forms	5
128	4568	Printer ribbons	10
535	4568	Printer ribbons	10
	4567	Computer forms	20
	4555	A4 stationery	45
	4550	Size1 pads	25

Figure 3.2(d) Data structures in 1NF.

Thus to convert ORDER-REPORT of Figure 3.2(c) into 1NF, we remove the four repeating elements and group them into a new table ORDER record, as shown in Figure 3.2(d). The rest Product No, Product Description, and Quantity-ordered are grouped into second table ORDER-ITEM. To maintain the relationship with the ORDER record, we make Order No part of the key with Product No for ORDER-ITEM.

Now the relation ORDER-ITEM is in first normal form, but it is not in its ideal form. Some of the attributes are not functionally dependent on the **primary key**, Order No and Product No. Some of the non-key attributes are dependent only on Product No, and not on the concatenated key.

Next we see how data structures in Figure 3.2(d) are converted to 2NF.

Second Normal Form (2NF)

A record is in 2NF if it is in 1NF and every data element is functionally dependent on both elements of non-key. As mentioned earlier, 2NF apply only to records that have compound keys.

ORDER-RECORD

Order No	Order Date	Invoice No	Invoice Date
25	15/8/97	1568	28/8/97
128	16/8/97	1569	28/8/97
535	20/8/97	1570	28/8/97

ORDER-ITEM

Order No	Product No	Quantity Ordered
25	4567	5
128	4568	10
535	4568	10
	4567	20
	4555	45
	4550	25

PRODUCT

Product No	Product Description
4567	Computer forms
4568	Printer ribbons
4555	A4 stationery
4550	Size1 pads

Figure 3.2(e) Data structures in 2NF.

Thus in Figure 3.2(d), in ORDER-ITEM table, Product description is not fully dependent on **composite key** of Order No + Product No. In fact Product Description is dependent only on Product No. To convert ORDER-ITEM into 2NF, we remove Product description and group into a separate table PRODUCT, with Product No as the key as shown in Figure 3.2(e).

Finally we normalize the data structure in Figure 3.2(e) into 3NF.

Third Normal Form (3NF)

A record is in 3NF, first if it is in 1NF and 2NF and secondly every data element is functionally dependent on the key. Also all non-key fields in data structure are independent of one another. If any data element is functionally dependent on another data element that is not the key of record, the record is not in 3NF.

ORDER-RECORD

Order No	Order Date	Invoice No
25	15/8/97	1568
128	16/8/97	1569
535	20/8/97	1570

ORDER-ITEM

Order No	Product No	Quantity Ordered
25	4567	5
128	4568	10
535	4568	10
535	4567	20
535	4555	45
535	4550	25

PRODUCT

Product No	Product Description
4567	Computer forms
4568	Printer ribbons
4555	A4 stationery
4550	Size1 pads

INVOICE

Invoice No	Invoice Date
1568	15/8/97
1569	16/8/97
1570	20/8/97

Figure 3.2(f) Data structures in 3NF.

In Figure 3.2(e), in ORDER table, we have Invoice date is dependent on both Order No and Invoice No. To convert the ORDER record into 3NF, we remove *Invoice Date* and group it into a new INVOICE record with Invoice No as the key as shown in Figure 3.2(f).

The above example is translated into functional dependency diagram as shown in Figure 3.2(g).

Conclusion. This way the ORDER-REPORT was transformed into the following relations:

ORDER-RECORD =   Order No
                              Order Date
                              Invoice No
ORDER-ITEM =   Order No
                              Product No
                              Quantity Ordered

56   Workbook on Systems Analysis and Design

```
┌─────────────────────────────────────────────────┐
│         ┌───────────┐      ┌──────────────┐     │
│         │ Invoice No│─────▶│ Invoice Date │     │
│         └───────────┘      └──────────────┘     │
└─────────────────────────────────────────────────┘
              ▲
┌─────────────────────────────────────────────────┐
│         ┌───────────┐      ┌──────────────┐     │
│         │ Order No  │─────▶│  Order Date  │     │
│         └───────────┘      └──────────────┘     │
│                                                 │
│         ┌───────────┐      ┌──────────────┐     │
│         │Product No │─────▶│   Product    │     │
│         └───────────┘      │ Description  │     │
│              │             └──────────────┘     │
└──────────────┼──────────────────────────────────┘
               ▼
         ┌───────────┐
         │ Quantity  │
         │ Ordered   │
         └───────────┘
```

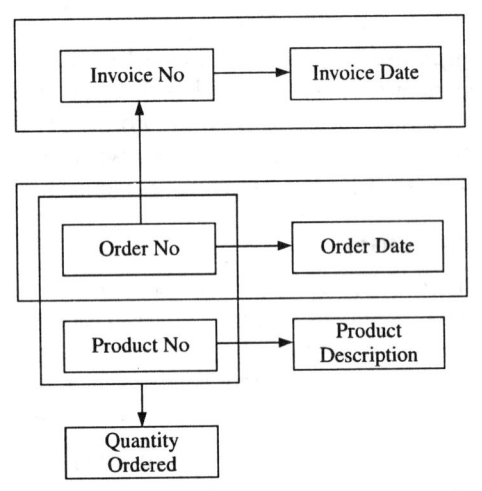

Figure 3.2(g)   Final dependency diagram for order entry system.

PRODUCT =   <u>Product No</u>
            Product Description

INVOICE =   <u>Invoice No</u>
            Invoice Date

            The data elements underlined are key fields.

Example 2

SALES-REPORT (Figure 3.2(h)) is an un-normalized relation since it has repeating groups. It is also important to observe that a single attribute such as *Salesperson-Number* cannot serve as a key to all the other data elements. This is clear when we examine the relationship between *Salesperson-No* and the other elements. Although there is one-to-one correspondence between Salesperson-No and two attributes (Salesperson-Name and Sales-Area), there is one-to-many relationship between Salesperson-No and the rest of attributes (Customer-No, Customer-Name, Warehouse-No, Warehouse-Location, and Sales-Amount).

Sales-person No	Sales-person Name	Sales Area	Customer No	Customer Name	Warehouse No	Warehouse Location	Sales Amount (Rs)
501	Bhushan	West	765	CMS	4	Bombay	134500
			760	IBM–AG	3	Turbhe	255700
			455	Lovleick	4	Bombay	457500
345	Vik Pradhan	South	1645	Infosys	2	Bangalore	55500
			247	Microland	1	Madras	125000
Etc.							

Figure 3.2(h)   Sales report.

Solution

Sales-person No	Sales-person Name	Sales Area	Customer No	Customer Name	Warehouse No	Warehouse Location	Sales Amount (Rs)
501	Bhushan	West	765	CMS	4	Bombay	134500
501	Bhushan	West	760	IBM–AG	3	Turbhe	255700
501	Bhushan	West	455	Lovleick	4	Bombay	457500
345	Vik Pradhan	South	1645	Infosys	2	Bangalore	55500
345	Vik Pradhan	South	247	Microland	1	Madras	125000
Etc.							

Figure 3.2(i) Normalized sales report.

From the figure, it is amply clear that the data elements for SALES-REPORT are: Salesperson-No, Salesperson Name, Sales Area, Customer No, Customer Name, Warehouse No, Warehouse Location, Sales Amount.

The normalized sales report Figure 3.2(i) is next converted into 1NF.

First Normal Form (1NF)

A record is in first normal form (1NF) if there are no data elements that repeat a variable number of times. The un-normalized relation SALES-REPORT is normalized into two separate structures. These new structures will be named SALESPERSON and SALESPERSON-CUSTOMER as shown in Figure 3.2j.

SALESPERSON

Salesperson No	Salesperson Name	Sales Area
501	Bhushan	West
345	Vik Pradhan	South

SALESPERSON-CUSTOMER

Salesperson No	Customer No	Customer Name	Warehouse No	Warehouse Location	Sales Amount (Rs)
501	765	CMS	4	Bombay	134500
501	760	IBM–AG	3	Turbhe	255700
501	455	Lovleick	4	Bombay	457500
345	1645	Infosys	2	Bangalore	55500
345	247	Microland	1	Madras	125000

Figure 3.2(j) Data structures in 1NF.

The record SALESPERSON contains the *primary key Salesperson No* and the other attributes that are not repeating, *Salesperson Name and Sales area.*

The second record SALESPERSON-CUSTOMER contains the primary key from the record SALESPERSON, as well as all of the remaining attributes that were part of repeating group. The record SALESPERSON-CUSTOMER is a first normal form structure, but it is not in its ideal form. Some of the attributes are not functionally dependent on the *primary key, Salesperson No and Customer No.*

Second Normal Form (2NF)

In the second normal form all of the attributes will be functionally dependent on the primary key. In the Figure 3.2(k) the structure SALESPERSON-CUSTOMER of 1NF is split into two new records; SALES and CUSTOMER-WAREHOUSE.

SALESPERSON

Salesperson No	Salesperson Name	Sales Area
501	Bhushan	West
345	Badrinarayan	South

SALES

Salesperson No	Customer No	Sales Amount
501	765	134500
501	760	255700
501	455	457500
345	1645	55500
345	247	125000

CUSTOMER-WAREHOUSE

Customer No	Customer Name	Warehouse No	Warehouse Location
765	CMS	4	Bombay
760	IBM–AG	3	Turbhe
455	Lovleick	4	Bombay
1645	Infosys	2	Bangalore
247	Microland	1	Madras

Figure 3.2(k) Data structures in 2NF.

The structure CUSTOMER-WAREHOUSE is in 2NF. It can be still normalized because there are additional dependencies within the relation. The non-key attributes are dependent not only on the *primary key Customer No*, but also on a *non-key* attribute, *Warehouse No*, which is against the principle of 2NF.

The *primary key* for the SALES records will be composite key of *Salesperson No.* and *Customer No.*

Third Normal Form (3NF)

A record is in 3NF, first it is in 1NF and 2NF and secondly every data element is functionally dependent on the key. In other words a record is in 3NF if all the non-key attributes are fully functionally dependent on the primary key and there are no non-key dependencies.

SALESPERSON

Salesperson No	Salesperson Name	Sales Area
501	Bhushan	West
345	Badrinarayan	South

SALES

Salesperson No	Customer No	Sales Amount
501	765	134500
501	760	255700
501	455	457500
345	1645	55500
345	247	125000

CUSTOMER

Customer No	Customer Name	Warehouse No
765	CMS	4
760	IBM–AG	3
455	Lovleick	4
1645	Infosys	2
247	Microland	1

WAREHOUSE

Warehouse No	Warehouse Location
4	Bombay
3	Turbhe
2	Bangalore
1	Madras

Figure 3.2(l) Data structure in 3NF.

Thus CUSTOMER-WAREHOUSE is normalized into two structures, CUSTOMER and WAREHOUSE. The primary key for the WAREHOUSE record will be Warehouse No.

Finally, the original un-normalized relation SALES-REPORT has been transformed into four 3NF structures.

SALESPERSON   Salesperson No, Salesperson Name, Sales Area
SALES   Salesperson No, Customer No, Sales Amount
CUSTOMER   Customer No, Customer Name, Warehouse No
WAREHOUSE   Warehouse No, Warehouse Location

The data elements underlined are the key fields.

Exercise

Shown below is a final dependency diagram. The system deals with employee time sheets. Work out the normalized tables from the Figure 3.2(m).

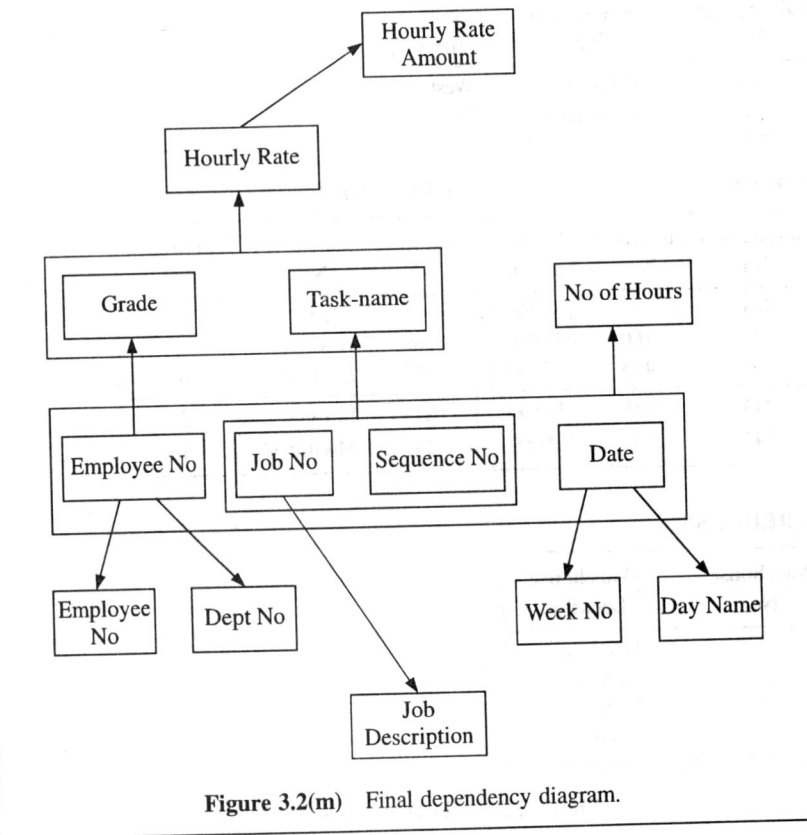

Figure 3.2(m) Final dependency diagram.

Solution to Exercise

Table Name	Data Elements
Employee	<u>Employee No</u>, Employee Name, Grade, Dept No
Job	<u>Job No</u>, Job Description
Task	<u>Job No, Sequence No</u>, Task Name
Day	<u>Date</u>, Week No, Day Name
Work	<u>Employee No, Job No, Sequence No, Date</u>, No. of Hours
Hour	<u>Grade, Task Name</u>, Hourly Rate
Rate	<u>Hourly Rate</u>, Hourly Rate Amount

The data elements underlined are the key fields.

Effective Table Design

No matter how well designed your database is, poor table design will lead to poor performance. Not only that, but overly rigid adherence to relational table designs will lead to poor performance. This is due to the fact that while fully relational table designs, said to be in the Third Normal form, are logically desirable, they are physically undesirable.

The problem with such designs is that although they accurately reflect the ways in which an application's data is related to one another, they do not reflect the normal access paths that users will employ to access that data.

> *Example*
>
> When you run a statement for querying data from tables, which returns large number of column's scattered among several tables, then the tables need to be joined. And if one of the joined tables is large, then the performance of the whole query may suffer.

Hence in designing the tables for an application, designers should therefore consider de-normalizing data. Hence it's advocated to create small summary tables from large, static tables.

The question is, if the data can be dynamically derived from the large, static tables on demand by end users. The answer is "Yes". But if the users frequently request it, and the data is largely unchanging, then it makes sense to periodically store that data in the format in which the users will ask for it.

The bottom line is, emphasis should be on providing the users the most direct path possible to the data in the format they want. User-centred table design, rather than theory-centred table design, will yield a system that meets the users requirement better.

Logical Modeling Conventions and Optimization

The data structures, namely tables in RDBMS terms can be related to each other via, three types of Entity-Relationships, namely **one-to-one (1:1), one-to-many (1:M) and many-to-many (M:M) relationships**. We will see how these relationships will help in building logical data structures.

A **primary key** (PK) is the column (field) or set of columns that makes each record in a table unique. A **foreign key** (FK) is a set of columns that refers back to an existing primary key. Tables can be related to each other via these three types of relationships.

- In a 1:1 relationship, the tables share a common primary key.
- In a 1:M relationship, a single record in one table is related to many records in another table.

62 Workbook on Systems Analysis and Design

- In M:M relationships, many records in one table are related to many records in another table.

Example

1:1 Relationships

Consider the table AC_MAST with the following columns.

AC_TYPE (PK)	AC_NO (PK)	AC_TITLE	AC_NAME
SB	1000551	MR.	Madhukar K
....	....		

Figure 3.2(n) AC_MAST table

If we need to add few columns for Addresses to the AC_MAST table, it will force the database to read through all the Address values every time the table is queried, even if only the AC_NAME field is being sought.

Hence to improve performance, create a second auxiliary table to AC_MAST, namely AC_MAST_ADD, shown in Figure 3.2(o).

AC_TYPE (PK)	AC_NO (PK)	AC_ADDR_1	AC_ADDR_2
SB	1000551	Plot No. 64, Cybertech House	J.B.Sawant, Thane
....	....		

Figure 3.2(o) AC_MAST_ADD table

The two tables AC_MAST and AC_MAST_ADD thus have 1:1 relationships as depicted in Figure 3.2(p).

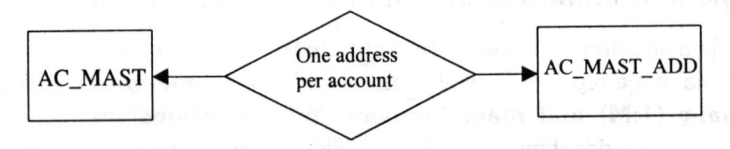

Figure 3.2 (p) E-R diagram: 1:1 relationships.

Example

1:M relationships

Consider the table AC_MAST in Figure 3.2(n). Here there will be one customer

Database Design **63**

no allocated for every type of accounts (namely Savings, Currents, Fixed Deposits, etc.), however one AC type may have various customer accounts. In this case a new table, AC_TYPES, Figure 3.2.(q), can be created. The AC_TYPE column of the AC_MAST table would be a foreign key to this new table.

AC_TYPE (PK)	AC_DESC
SB	Savings Bank
CD	Current Account
.....	.....

Figure 3.2(q) AC_TYPES table.

The two tables AC_MAST and AC_MAST_ADD thus have 1:1 relationships as depicted in Figure 3.2(r).

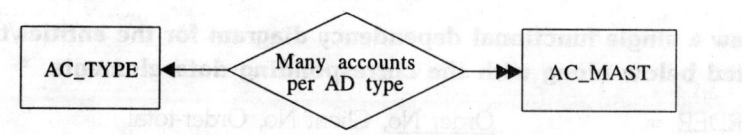

AC_TYPE ◄── Many accounts per AD type ──► AC_MAST

Figure 3.2(r) E-R diagram: 1:M relationships.

Example

M:M relationships

It may be possible that many records (rows) of a table are related to many rows of another table. Looking further to AC_TYPES table in Figure 3.2(q), for each account types there can be many customers in CUST_MAST table. Similarly one customer can open accounts of various account types.

CUST_NO (PK)	CUST_NAME
G2051	Moose & Rose
S1055	Smile & Tiles
.....	.....

Figure 3.2(s) CUST_MAST table.

The two tables AC_MAST and AC_MAST_ADD thus have 1:1 relationships as depicted in Figure 3.2(t).

```
         ┌─────────┐      ╱╲       ┌──────────┐
         │         │◄────╱  ╲────►►│          │
         │AC_TYPES │    ╱Many╲     │CUST_MAST │
         │         │    ╲Accts╱    │          │
         └─────────┘     ╲  ╱      └──────────┘
                          ╲╱
```

Figure 3.2(t)   E-R diagram: M:M relationships

Primary Key (PK) Constraints

Relational database theory dictates that one or more columns' in a table that can uniquely identify each row in that table is the object's primary key. Creating primary key for table's assist the application tuning process since primary key definitions in the data dictionary ensures uniqueness between rows in a table. Also data retrieval using entries stored in primary key is faster than using an index that does not contain unique values.

Practice for Section 3.2

#1
Draw a single functional dependency diagram for the entities/tables listed below along with the corresponding data elements.

ORDER =	<u>Order No</u>, Client No, Order-total
CLIENT =	<u>Client No</u>, Client-address, Credit-limit
ORDER-LINE =	<u>Line No</u>, <u>Order No</u>, <u>Part No</u>, <u>Sequence No</u>, Quantity, Line-total
PART =	<u>Part No</u>, Description, Price
STOCK =	<u>Part No</u>, Stock-quantity, Location

#2
A doctor manually maintains his patient's records as listed below. Produce a normalized design for the entities, which you would use. Also draw final functional dependency for the above normalized structure.

>   PATIENT (National-insurance-number, Patient Name, Sex, Address, Telephone-number, Date-of-last-visit, *(Contact, Next-of-kin-telephone-number, Relationship) *(Date-of-visit, Symptoms, Comments, Medication) *(Date-of-hospitalization, Hospital-name, Reason, date-left, Doctor attended))

>   * For each patient details, have 1 or more, Contacts of the relations, Visits for checkup, Hospitalization.

#3
In the share accounting system the structure of PENDING-ORDERS data store is shown below:

>   PENDING ORDERS
>   Customer-Name

Customer-Address
Phone
Orders * (1-)
    Company-Name
    Company-address
    Buy-or-sell
    Unit Price
    No-of-shares
    Date-of-order
    Total-value

Convert this data structure into first normal form. Further convert the 1NF structures into 2NF and then to 3NF.

#4
Shown below is the report structure of Parts supplied by various suppliers. Normalize the relation into different structures.

SUPPLIER-PART
    Supplier No
    Part No
    Part Description
    Supplier Name
    Supplier Address
    Quantity supplied

Assume there can be more than one supplier for any part.

#5
In the Materials Purchase and Billing system, the structure for printing INVOICE is shown below. **Normalize this structure into 1NF, 2NF and 3NF.**

INVOICE
    Invoice-No
    Order-No
    Invoice date
    Customer-No
    Customer-name
    Item-details * (n-m)
        Item No
        Item Description
        Quantity Sold
        Rate
        Discount
        Total value
    Invoice Amount

#6

A customer in a bank is sent a statement of his balance along with receipt/payment details. The data elements which forms part of the statement are:

CUSTOMER-STATEMENT
    Customer No
    Customer Name
    Address
    Last Statement Date
    Last Statement Balance
    Customer Credit Limit
    Statement Date
    Statement Balance
    Payment/Receipt Details *(1-)
        Date
        Amount
        {Payment}{Receipt}

Normalize the above structure into 1NF, 2NF, 3NF.

4
Input-Output Design

The documents, reports and screens of a system are the user interface. The user interface is the only part of the system that the user sees. The rest is invisible. The user interface, therefore is the most important part of the system to the user. Its design must be such that in addition to providing information to the user it should be appealing.

The steps in designing the interface for a system are:

1. Create the interface summary table for the system
   - **1.1.** Review the human-computer boundary in the DFD.
   - **1.2.** List the interfaces and review the design to define data elements to be included in each form/document, input screen and output report.
2. Design external input forms.
3. Design the human-computer dialogue
   - **3.1.** Select the type(s) of dialogue
   - **3.2.** Design each dialogue
4. Design input (data entry) screens.
5. Design reports
   - **5.1.** Select the output method for each report
   - **5.2.** Design internal reports
   - **5.3.** Design external reports

Figure 4.1(a) highlights the interface design process. The inputs to the design process are the user characteristics, the data flow diagram, and the design repository.

4.1 INPUT FORM DESIGN

The quality of system input determines the quality of system output. It is vital that input forms and screens, and output reports be designed with this critical relationship in mind.

```
            |              |
          User         Data flow
      Characteristics   diagram
            |              |
            ▼              ▼
      ┌─────────────────────────┐
      │         3.1             │
      │     Design User         │◄──────────────┐
      │     Interface           │               │
      └─────────────────────────┘               │
            |         |                         │
            ▼         ▼                 ┌───┬──────────────────┐
           Data    Interface            │D1 │ Design Repository│
           Entry   Specifications       └───┴──────────────────┘
          Controls
            |         |
            ▼         ▼
```

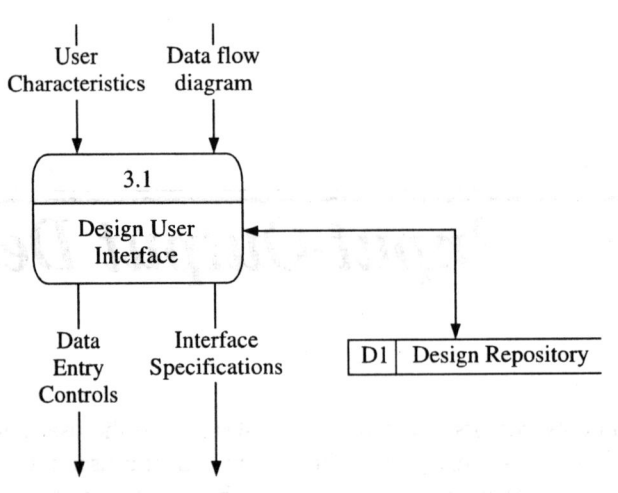

Figure 4.1(a) Inputs and Outputs for interface design.

Well-designed input forms and visual display screens should meet the objectives of effectiveness, accuracy, ease of use, consistency, simplicity, and attractiveness. All of these objectives are attainable through the use of basic design principles, knowledge of what is needed as input for the system, and an understanding of how users respond to different elements of forms and screens.

Effectiveness means that input forms and screens serve specific purpose in the management information system, while accuracy refers to design that assures proper completion. Ease of use means that forms and screens are straightforward and require no extra time to decipher.

Consistency in this case means that forms and screens group data similarly from one application to the next, while simplicity refers to keeping forms and screens purposely uncluttered in a manner that focuses the user's attention. Attractiveness implies that users will enjoy using, or even be drawn to using, forms and screens through their appealing design.

Objectives

Input Forms elicit and capture information required by organizational members that will often input to the computer. Through this process, forms often serve as source document for data entry personnel.

Four Guidelines for Form Design

- Make forms easy to fill out.
- Ensure that forms meet the purpose for which they are designed.
- Keep forms attractive.
- Design formats to assure accurate completion.

There are a number of means to achieve each guideline for form design. We will discuss the above four guidelines:

Input-Output Design   **69**

Making forms easy to fill out

Designing a form with proper flow can minimize the time and effort expended by employees in form completion. Forms should flow from left to right and top to bottom, as shown in the police incident report depicted in Figure 4.1(b).

The flow of incident report, as shown in Figure 4.1(b), works because it is based on the way people read a page. The incident report is designed so that the attending officer first fills in the date, then the time, then continues onto the bottom of the form, which elicits suggestions on further handling of the situation.

Another technique that makes it easy for people to fill out forms correctly is logical grouping of information. The main sections, into which they are grouped, as shown in Figure 4.1(c), are:

➢ Heading
  - Identification and access
  - Instructions
➢ Body
➢ Signature and verification
➢ Totals
➢ Comments

Typical Information flow

Figure 4.1(b) represents how a "Incident Report" is filled by the investigating officer.

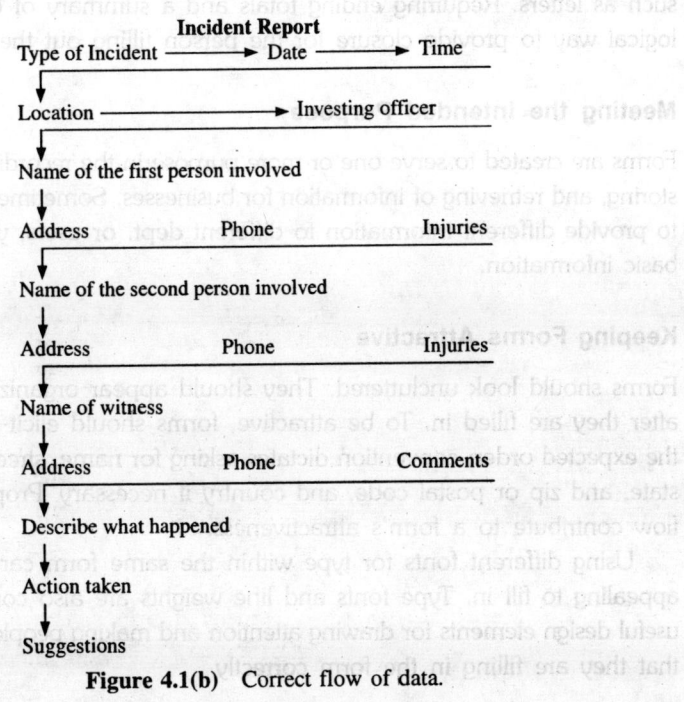

Incident Report

Type of Incident ⟶ Date ⟶ Time
↓
Location ⟶ Investing officer
↓
Name of the first person involved
↓
Address   Phone   Injuries
↓
Name of the second person involved
↓
Address   Phone   Injuries
↓
Name of witness
↓
Address   Phone   Comments
↓
Describe what happened
↓
Action taken
↓
Suggestions

Figure 4.1(b) Correct flow of data.

Well Designed Form

Figure 4.1(c) illustrates how a well designed form should be.

Heading	Identification and Access
Instructions	
Body	
Signature and Verification	Totals
Comments	

Figure 4.1(c) Sections found in well-designed form.

The heading section usually includes the name and address of the business originating the form. The identification and access section includes codes that may be used to identify the form and user accessing it.

The middle of the form is its body, which requires the most detail and explicit information about the process.

The bottom quarter of the form is composed of three sections: signature and verification, totals, and comments. By requiring a signature in this part of the form, the designer is echoing the design of other familiar documents, such as letters. Requiring ending totals and a summary of comments is a logical way to provide closure for the person filling out the form.

Meeting the Intended Purpose

Forms are created to serve one or more purpose in the recording, processing, storing, and retrieving of information for businesses. Sometimes it is desirable to provide different information to different dept. or users, yet share some basic information.

Keeping Forms Attractive

Forms should look uncluttered. They should appear organized and logical after they are filled in. To be attractive, forms should elicit information in the expected order; convention dictates asking for name, street address, city, state, and zip or postal code, and country if necessary. Proper layout and flow contribute to a form's attractiveness.

Using different fonts for type within the same form can help make it appealing to fill in. Type fonts and line weights are also considered to be useful design elements for drawing attention and making people feel confident that they are filling in the form correctly.

Example

The Solomon Brothers employee expense voucher, shown in Figure 4.1(d) goes a long way toward securing accurate form completion utilized in this sample expense voucher. The form design implements the correct flow: top to bottom, and left to right. It also observes the idea of seven main sections or information categories. Additionally, the employee expense voucher uses a combination of clear captions and instruction.

Since Solomon Brothers employees are reimbursed only for actual expenses, getting a correct total expenditure is essential. The form design provides an internal double check with column totals and row totals expected to sum to the same number. If the row and column totals don't sum to the same number, the employee filling out the form knows there is a problem and can correct it on the spot. An error is prevented, and the employee can be reimbursed the amount due; both outcomes are attributable to a suitable form design.

Figure 4.1(d), shows a sample format of form as per the need.

Designing formats to assure accurate completion

SOLOMON BROTHERS
EMPLOYEE EXPENSE VOUCHER

Social Security Number

Claimant: Make no Entries in Shaded Areas
Full Name of Employee_____

Voucher Number
Action Taken ON:

Department_____Room Number_____

List expenses for each day separately, attach receipts for all expenses except meals, taxis and miscellaneous items then Rs 30.00. Item-size all miscellaneous expenses.

Date 19__	Place, City, State	Meal Expenses	Lodging Expenses	Automobile		Miscellaneous		Taxi Cost	Total Cost
				Miles	Cost	Description	Cost		
Totals									

I certify that all the above information is correct

Signature of Claimant _____ Date _____

Approved by _____ Date _____

Form BB-104 8.85

Figure 4.1(d) Sample document format.

Practice for Section 4.1

#1

The Millennium Bank has gone for automation of their front office operations. In the process, they have to change their current manual receipts and payment vouchers, which the customers use for withdrawal and deposits of cash/cheque/demand drafts.

The information, which these forms must capture for the new computer system is:

Deposit Form

1. Date
2. Type of Accounts: Savings/Current/Loans, etc.
3. Account No of Customer
4. Name of Customer
5. Instrument type: Cash/Cheque etc.
6. If Cheque then, Amount, Cheque No, Cheque Date, Bank and Branch Name
7. If Cash, then provision for denominations type, nos. and total for each denomination
8. Total Deposit amount, in figures and words
9. Signature of customer
10. ID of the Cashier accepting Cash/Cheque

Make provisions in a single deposit form, to accept upto 5 Cheques.

Withdrawal Form

1. Date of Withdrawal
2. Type of Accounts: Savings/Current/Loans, etc.
3. Account Number of Customer
4. Name of Customer
5. Amount in figures and words
6. Provision for denomination type, number and total for each denomination. The cashier will enter this at the time of payment
7. Signature of Customer
8. ID of Officer (optional) who authorized the payment
9. ID of cashier making payment

Based on the above description, do the following

Design the form for Deposits and Payments. The form must include the Bank Logo and Name, Branch Code and Name.

#2

Get Well Clinic is in the process of computerizing the clinics operations, including the doctor's prescription form. This prescription form will be used by the chemist store attached to the clinic.

Design a form to include the following information:
1. Date and Patient Admission Number
2. Name of Patient
3. Sex
4. Age
5. Illness Type
6. Medicine Name
7. Daily Dosage
8. No of Days
9. Doctors Name & ID

The information 6, 7 and 8 must be captured in tabular form.

74 *Workbook on Systems Analysis and Design*

4.2 INPUT SCREEN DESIGN

A data entry screen is used to enter data into computer files from source documents (input forms) or reports, as well as to update data.

The four guidelines for screen design are important but not exhaustive. They include:

- Keep the screen simple
- Keep the screen presentation consistent
- Facilitate user movement among screens
- Create an attractive screen

This will lead to two issues important in designing data entry screens:

1. Speed of data entry
2. Accuracy of data entry

How good a user interface is, depends very much on the user of the interface.

Example

A data entry user who has to enter 100 or more purchase orders every day will care most about the speed of data entry. A usable interface in such case would require a minimum number of keystrokes to enter the order.

Likewise, a company director, who needs to check on the status of a project once or twice a week, will care most about the ease with which the system can be used, and the way information is presented to him.

Keeping the Screen Simple

The first guideline for good screen design is to keep the screen display simple. The display screen should show only that which is necessary for the particular action being undertaken. Figure 4.2(a) shows a division of the screen into 3 sections, useful in simplifying interactions with screens. The top of the screen features a "heading" section, to describe to the user where he or she is in the package.

HEADING (and KEY Words for PULL-DOWN menus)

BODY

(Use the conventions for reading left to right and top-to-bottom in order to make it easy for the user to enter the data.

COMMENTS AND INSTRUCTIONS FOR FUNCTION KEYS

Figure 4.2(a) Divisions of data entry screen.

The middle section is called the "body" of the screen. This can be used for data entry and is organized from left to right and top to bottom. Field definitions showing how many data are allowable in each field of the screen's body should also be supplied to the user. This can be accomplished by using **brackets, braces, or ampersands** to denote the start and end of a field.

The third section of the screen is the "Comments and Instructions" section. This section may display a short menu of commands that remind the user of basics such as how to change screens or functions, save the file, or terminate entry.

Inclusion of such basics can make inexperienced users feel infinitely more secure about their ability to operate the computer without causing a fatal error. Instructions in the third section could also list acceptable code choices the user needs for completing data entry.

Example

Figure 4.2(b) shows a sample order entry input screen based on the divisions of data entry screen as explained above.

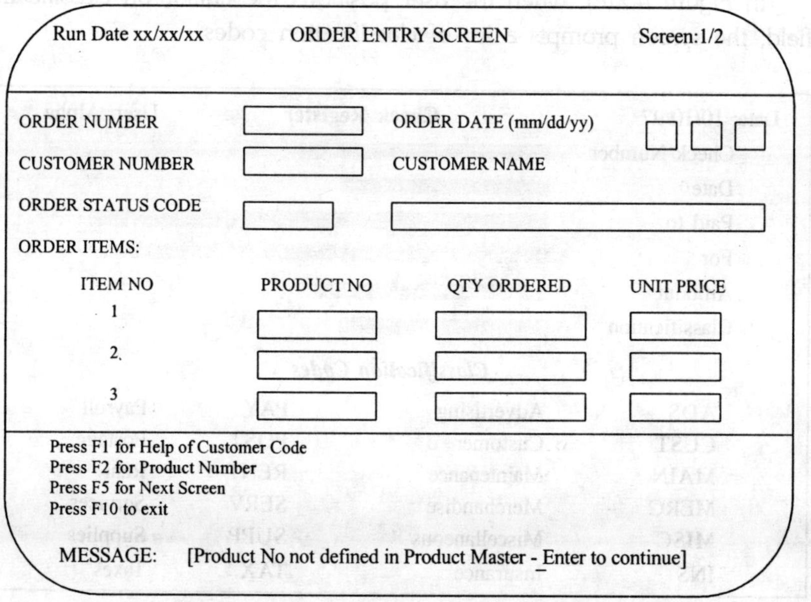

```
   Run Date xx/xx/xx         ORDER ENTRY SCREEN            Screen: 1/2


   ORDER NUMBER        [        ]   ORDER DATE (mm/dd/yy)   [ ][ ][ ]
   CUSTOMER NUMBER     [        ]   CUSTOMER NAME
   ORDER STATUS CODE   [        ]
   ORDER ITEMS:
          ITEM NO         PRODUCT NO        QTY ORDERED      UNIT PRICE
            1
            2.
            3

   Press F1 for Help of Customer Code
   Press F2 for Product Number
   Press F5 for Next Screen
   Press F10 to exit

      MESSAGE:   [Product No not defined in Product Master - Enter to continue]
```

Figure 4.2(b) Sample order entry input screen.

Keeping the Screen Consistent

The second guidelines for good screen design is to keep the screen display consistent. Screens can be kept consistent by locating information in the same area, each time a new screen is accessed. Also, information that logically belongs together should be consistently grouped together. For example, Name and address go together, not name and zip code.

Example

The title of the program, as shown in Figure 4.2(b) is always at the top and is centred, the instructions are always at the bottom.

Also the individual function keys used will serve the same purpose, throughout the applications (example, F10 to exit the application).

Another way to keep the screen display informative is to list a few basic commands that, when used will overlay windows to partially or totally fill the current screen with new information. Users can minimize or maximize the size of windows as needed. This is shown in Figure 4.2(c) and Figure 4.2(d).

Also the fields of the screen should be in an order consistent with that of the fields of the source form. In this way, users start with a simple, well-designed screen whose complexity they control through the use of multiple windows.

In Figure 4.2(c), when the user positions the cursor on classification field, the screen prompts a list of classification codes.

```
Date: 10/10/97           Check Register           User: Alpha
    Check Number    ▓▓▓▓▓▓▓▓▓▓▓▓
    Date            ▓▓▓▓▓▓▓▓▓▓▓▓
    Paid to         ▓▓▓▓▓▓▓▓▓▓▓▓▓▓▓▓▓▓▓▓▓▓▓▓▓▓▓
    For             ▓▓▓▓▓▓▓▓▓▓▓▓▓▓▓▓▓▓▓▓▓▓▓▓▓
    Amount          ▓▓▓▓▓▓▓▓▓▓▓▓
    Classification  ▓▓▓▓▓▓▓▓▓▓▓▓

              Classification Codes
    ADS    Advertising      PAY    Payroll
    CUST   Customer         POST   Postage
    MAIN   Maintenance      RENT   Rent
    MERC   Merchandise      SERV   Services
    MISC   Miscellaneous    SUPP   Supplies
    INS    Insurance        TAX    Taxes
```

Figure 4.2(c) Screen with classification codes at the bottom.

Figure 4.2(d) shows an alternate screen display for the program along with the window of classification codes the user has asked for. The format is designed such that by pressing F6 OR "?", the user is able to access the necessary codes.

Making this window available facilitates quick and correct entry since the user need not remember the infrequent used codes or leave the screen to check a list of them while entering the data, online.

Input-Output Design   **77**

```
Cheque Number  [_____]      Classification Codes
         Date  [_____]      ADS    Advertising
      Paid to  [_____]      CUST   Customer
          For  [_____]      MAIN   Maintenance
       Amount  [_____]      MERC   Merchandise
Classification [_____]      MISC   Miscellaneous
                                   INS    Insurance
                                   PAY    Payroll
                                   POST   Postage
                                   RENT   Rent
                                   SUPP   Supplies
                                   TAX    Taxes

Message [ Type F6 OR ? to get help of Classification Codes    ]
```

Figure 4.2(d)   Another way to represent codes.

Facilitating Movement between Screens

The third guideline for good screen design is to make it easy to move from one screen to another. One common method for movement is to have users feel as if they are physically moving to a new screen. This can be done by use of Scrolling, Calling another screen for more details, and through use of on-screen dialog.

Screens should be so designed that user is always aware of the status of an action. This can be done in following ways.

(a) Use error messages to provide feedback on mistakes

(b) Use confirmation messages to provide feedback on update actions

(c) Use status message, when some backend process is taking place

(d) Allow the user to reverse an action when possible, or query the user when actions are far-reaching, example; "Do you want to Delete entry (Y/N)".

Create Attractive Screen

The fourth guideline for good screen design is to create attractive screens for the user, but not a flashy one. If users find screens appealing, they are likely to be more productive, need less supervision, and make fewer errors.

Screens should draw the user into them and hold their attention. This is accomplished by the use of plenty of open area surrounding data-entry fields, so the screen achieves an uncluttered appearance as shown in Figure 4.2(b).

Additional Tips for Data/Text I-O while Designing Screens

➤ Place a descriptive title on each screen
➤ Indicate on each screen how to exit from the screen
➤ Avoid hyphenation of words between lines
➤ Use abbreviations only when the user easily understands them
➤ All formats should be in a form that is familiar to the users example alphabetic data should be left justified, and so on.
➤ A message should contain enough information to correct a problem without referring to additional documentation.

Graphical User Interface Design

A graphical user interface (GUI) uses a Windows or other similar graphics screen for entering and displaying data. While these screens have traditional data entry and display fields, several additional features are also included in the screen design. These are illustrated in Figure 4.2(e).

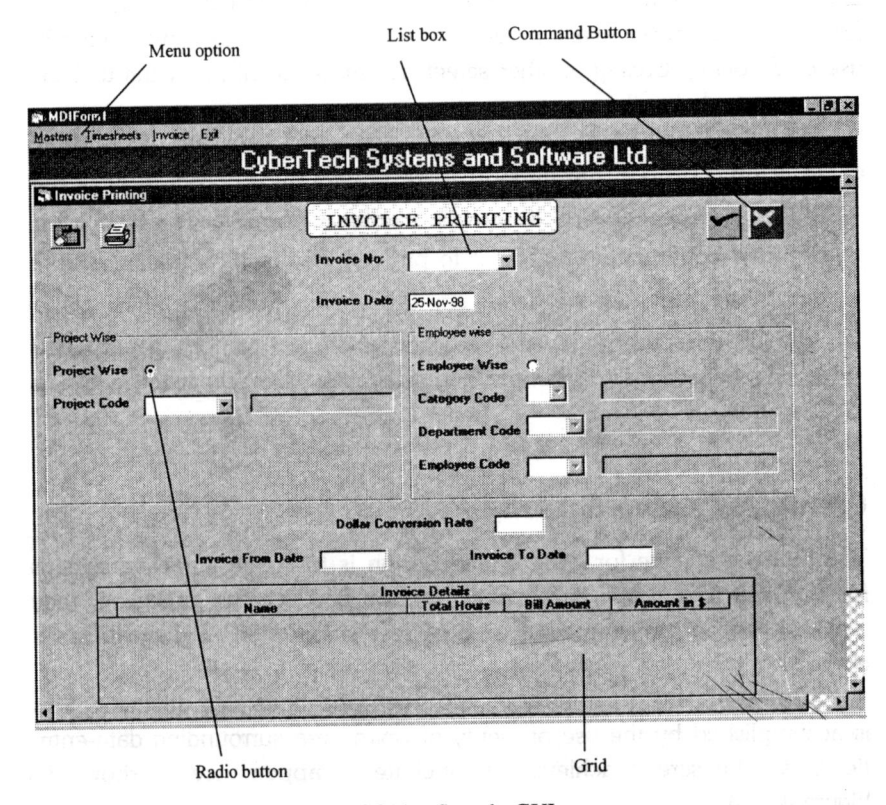

Figure 4.2(e) Sample GUI screen.

Input-Output Design

The screen shown is used to add a customer order and has been designed to illustrate many of the GUI features.

Example

> A CHECK BOX is used to indicate a new customer, Check boxes contain a data or are empty, corresponding to whether or not the user selects the option.

> A circle, called RADIO BUTTON, is used to select exclusive choices: either of the options can be chosen, but not both.

> A LIST BOX is one, which displays several options that may be selected with the mouse. A DROP-DOWN list box is used when there is little room available on the screen.

> A COMMAND button performs an action when the user selects it with the mouse.

Practice for Section 4.2

#1

At branches of Millennium Bank, the customers have to fill a manual form each for Cash Deposits and Cash Withdrawal. The information, which these forms capture are:

Cash Deposit Form

1. Date
2. Type of Accounts: Savings/Current/Loans, etc.
3. Account No of Customer
4. Name of Customer
5. Provision for cash denominations type, nos. and total for each denomination
6. Total Deposit amount, in figures and words.
7. Signature of customer
8. ID of the Cashier accepting Cash

The deposit form has provisions to accept multiple denominations.

Withdrawal Form

1. Date of Withdrawal
2. Type of Accounts: Savings/Current/Loans, etc.
3. Account No of Customer
4. Name of Customer
5. Amount in figures and words
6. Provision for denomination type, nos. and total amount for each denomination—to be entered by Customer at the time of payment.
7. Signature of Customer
8. ID of Officer (optional) who authorized the payment.
9. ID of Cashier making payment.

Based on the above description:

(a) **Design a Character based Input Entry Screen for the above process.**

(b) **Also design the Input screen for GUI, using Check box, Radio button, List box, Data Grid, etc.**

4.3 MENU DESIGN

The human-computer dialogue defines how the interaction between the user and the computer takes place. There are two common types of dialogues:

1. Menu
2. Question and Answer.

With a menu dialogue, a menu displays a list of alternate selections. The user makes a selection by choosing the number or letter of the desired alternative. Menu dialogues are of the most common type, because they are appropriate for both frequent and infrequent users of a system.

```
          ORDER TRACKING MENU
    Options:
      1. Enter Orders
      2. Enter Invoices
      3. Print Order Register Report
      4. Exit to Main Menu
      5. Help for Order Tracking Process

      Enter Your Option:   5
```

Figure 4.3(a) Good menu screen.

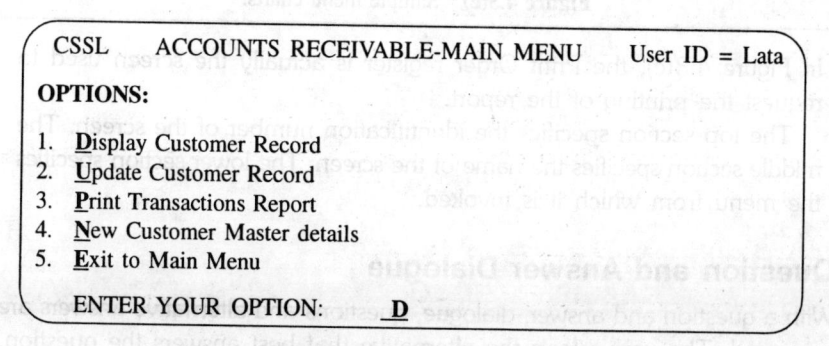

```
   CSSL      ACCOUNTS RECEIVABLE-MAIN MENU      User ID = Lata
   OPTIONS:
     1. Display Customer Record
     2. Update Customer Record
     3. Print Transactions Report
     4. New Customer Master details
     5. Exit to Main Menu

     ENTER YOUR OPTION:        D
```

Figure 4.3(b) Alternate way of designing Menu.

Comparing Figure 4.3(a) and Figure 4.3(b), it is distinct that, apart from selecting option through number's, the user can enter the highlighted word to get into the process.

In menu selection, the users read a list of items and select the one most appropriate to their task, usually by highlighting the selection and pressing the return key, or by keying the menu item number.

Menu selection systems require a very complete and accurate analysis of user tasks to ensure that all the necessary functions are supported with clear and consistent terminology.

Conventions for Menu Charts

Menu/Dialogue charts should be created to show the relationship among menu options. These charts document how the user moves through menus, question and answer dialogues, data entry screens and output reports.

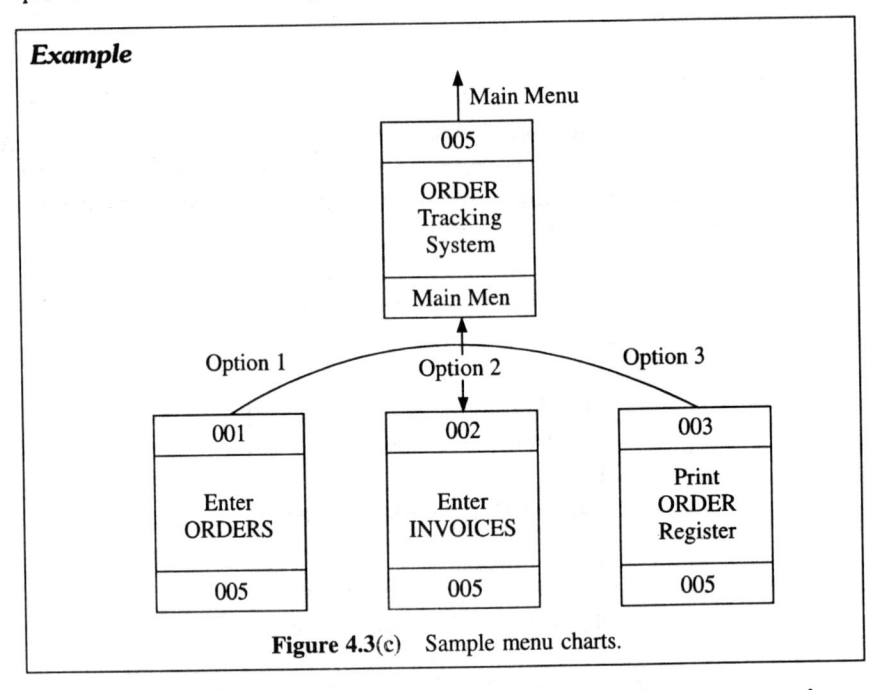

Example

```
                    ↑ Main Menu
                  ┌──────────┐
                  │   005    │
                  ├──────────┤
                  │  ORDER   │
                  │ Tracking │
                  │  System  │
                  ├──────────┤
                  │ Main Men │
                  └──────────┘
                        ↑
        Option 1      Option 2        Option 3
                        ↓
  ┌──────────┐   ┌──────────┐   ┌──────────┐
  │   001    │   │   002    │   │   003    │
  ├──────────┤   ├──────────┤   ├──────────┤
  │          │   │          │   │  Print   │
  │  Enter   │   │  Enter   │   │  ORDER   │
  │  ORDERS  │   │ INVOICES │   │ Register │
  ├──────────┤   ├──────────┤   ├──────────┤
  │   005    │   │   005    │   │   005    │
  └──────────┘   └──────────┘   └──────────┘
```

Figure 4.3(c) Sample menu charts.

In Figure 4.3(c), the Print Order register is actually the screen used to request the printing of the report.

The top section specifies the identification number of the screen. The middle section specifies the name of the screen. The lower section specifies the menu from which it is invoked.

Question and Answer Dialogue

With a question and answer, dialogue, questions and alternative answers are presented. The user selects the alternative that best answers the question. Question and answer dialogues are most appropriate for intermediate (between frequent and infrequent) users of a system

Example

After an order was entered on the Order Entry screen, the system might ask if user would like to create an Invoice for the entered order. To this the user can offer two answers – Yes OR No.

Do you want to enter Invoice (Y/N)?

Do you want to continue, enter more Orders (Y/N)?

4.4 OUTPUT DESIGN

Output is information delivered to users through information system. Output can take many forms:

- Traditional hard copy of printed reports
- Soft copy such as display screens
- Microforms
- Audio/Video output.

Again, the reports are classified into two types:

1. **Internal reports** which supply information to personnel within the organization. For example, Statistical reports, Order register report and so on.
2. **External reports** which may be given to customers/users outside the organization. Example, Statement of Accounts to customer in banks, Invoice to be sent and so on.

Since useful output is essential to ensuring use and acceptance of the information system, there are several objectives that the system analyst tries to attain when designing output.

There are five (5) objectives for output:

1. Design output to serve the intended purpose.
2. Design output to fit the user.
3. Deliver the appropriate quantity of output.
4. Assure that the output is where it is needed.
5. Provide the output on time.

We will discuss in short each of these objectives.

Designing output to serve the intended purpose

All output should have a purpose. It is not enough that a report or screen is made available to users because it is technologically possible to do so. During the information requirement determination phase of analysis, the systems analyst finds out what purposes must be served. Output is then designed based on those purposes.

Designing output to fit the user

With a large information system serving many users for many different purposes, it is difficult to personalize output for a specific user. However, it is possible to design output to fit a user function in the organization.

Delivering the appropriate quantity of output

It is difficult to decide what quantity of output is correct for the users since information requirements are in continuing flux. At the same time no one is served if excess information is given only to flaunt the capabilities of the system.

Always keep the decision-makers in mind when deciding about quantity of output. Often they will not need great amount of output, especially if there is an easy way to access more.

Making sure the output is where it is needed

Output is printed on paper, displayed on screens, piped over speakers and stored on microforms. Output is often produced at one location (for example, in the data processing department) and distributed to the user.

To be used and useful, output must be presented to the right user. No matter how well designed reports are, if they are not seen by the concerned decision-makers, they have no value.

Providing the output on time

One of the most common complaints of the users is that they do not receive information in time to make necessary decisions. Not only you as a designer have to be conscientious about which user is receiving what output, you must also be concerned about the timing of the output distribution.

Many reports are required on a daily basis, some only monthly, others annually, and still others, only by exception. Accurate timings of output can be critical to business operations.

Another important factor is the *response time*—the acceptable printing time for the report.

Steps in preparing the printer layout worksheet

The following are the steps for preparing the printer layout worksheet:

1. Determine the need for the report.
2. Determine the users.
3. Determine data items to be included.
4. Estimate the number of spaces necessary and decide on the overall size of the report.
5. Title the report.
6. Number the pages of the report.
7. Include the preparation date on the report.
8. Label each column of data appropriately.
9. Define the detail line for variable data by indicating whether each space is to be used for an alphabetic, special or numeric character.
10. Indicate the positioning of summaries (control breaks).
11. Review prototype reports with users and programmers for feasibility, usefulness, readability, understandability, and aesthetic appeal.

Practice for Section 4.4

#1

The following process specification was listed for Travel Arrangement System for "Taj Tours and Travels" The analyst has listed out the process modules for the system, describing the activity associated.

(i) The travel agency keeps a record of each customer. CUSTOMER-NO is a unique identifier of the CUSTOMER-NAME, and each customer has a CUSTOMER-ADDRESS and CONTACT-NO.

(ii) Each trip is identified by a unique TRIP-NO. Each customer can make any number of trips, but each trip is for only one customer. The DATE-ARRANGED, PLACES-TO-BE-VISITED and COST are stored for each trip.

(iii) Trips may be either PACKAGED-TRIPS or SPECIALLY-ARRANGED-TRIPS for group are people/corporate. These trips are undertaken by either Buses or Airs. The travel agency has its own Luxury/AC buses through which they conduct the tours.

(iv) Each packaged trip is for one packaged tour. A field called EXTRAS is provided for each packaged trip to record any special needs of the CUSTOMER.

(v) A specially arranged trip is one where the travel agency constructs a trip out of a number of bookings. The booking can be either a hotel booking or an airline booking. Each booking is given a unique LEG-NO within a TRIP-NO, and has FIXED-NO-OF-BOOKINGS. A special description is stored for each specially arranged trip.

(vi) A hotel reservation is made with one hotel. The travel agency has its own list of schedule hotels. Each hotel is identified by a unique HOTEL-ID, HOTEL-NAME and has an ADDRESS, CITY, TELEPHONE-NO and FAX-NO. The booking involves ROOM-TYPE, DAILY-RATE, DATE-IN and DATE-OUT.

(vii) An airline reservation is made with one airline. The travel agency has list of airlines with which they operate tours. Each AIRLINE-NAME is identified by a unique AIRLINE-CODE. Each airline reservation made is identified by RESERVATION-NO, and includes the DEPARTURE-AIRPORT and DEPARTURE-TIME. The reservation also includes details of all stopovers, of the ARRIVAL-TIME, DEPARTURE-DATE, DEPARTURE-TIME and AIRPORT.

Based on the above description do the following:

(a) Design a Standard Input Screen format, as discussed in the format—header, body and comments/Message screen split.

(b) Design all the necessary Input screens for the above processes. Assume all other information, which must be part of data structures to facilitate input/output data.

(c) **Also design a Menu to execute all the processes. The menu must categorize between Master and Transaction process and various MIS reports.**

#2

Shown below is an Interface table for pharmacy system for Bombay Hospital. For each interface is associated a type of design and the data elements to be associated.

Draw the interface design for the same. Assume your own data types and if to be Input or Output or Both.

No.	Name	Type	Data Elements
(i)	Doctor Prescription Entry Screen	Data Entry	Patient-ID and Name, Physician-ID and Name, Prescription No, Prescription Date, Prescribe-Drug Name, Drug-Quantity, Dosage.
(ii)	Prescription Update Screen	Data Entry	Prescription No and Date, Physician ID and Name, Patient ID and Name.
		Data Entry	* Prescribe-Drug Name, Unit Price, Drug-Quantity, Total Price, Dosage.
			Total Amount
			The * fields may be multiple for each prescription slip.
(iii)	Daily Drug Log	Internal report	Prescription No, Physician ID and Name, Patient ID and Name, Drug Name, Drug quantity and Dosage.
			* The report must be either separately for each Physician or a combined one for all physicians.
(iv)	Prescription Label (Printed from Prescription Update Screen)	External report To Patient	Prescription No, Physician ID and Name, Patient ID and Name.
			* Prescribe-Drug Name, Unit Price, Drug-Quantity, Total Price, Dosage.
			Total Amount
			The * fields may be multiple for each prescription slip.

5
Program Design

5.1 INTRODUCTION TO STRUCTURED PROGRAM DESIGN

Structured program design is a set of techniques, guidelines, and a method for making program coding, testing, and maintenance easier by reducing the complexity of programs. Structured design reduces the complexity by breaking the programs into small pieces called modules.

The techniques used for structured program design are the **structure chart** and **program Definition language (PDL)**. The structure chart allows us to create an outline of a program by specifying modules and how the modules are connected. Whereas PDL is used to specify the processing of each module, by way of Structured English, Decision trees and Decision tables.

```
┌─────────────────────────┐
│   Data flow diagram     │
│   User Procedures       │
│ Interface Specifications│
│ Data base specifications│
└─────────────────────────┘
            │
            ▼
    ┌──────────────┐
    │  Structured  │
    │   Program    │
    │    Design    │
    └──────────────┘
            │
            ▼
┌─────────────────────────┐
│ Updated Data dictionary │
│   Structured Charts     │
│          PDL            │
│ Coding & Testing strategy│
└─────────────────────────┘
            │
            ▼
    ┌──────────────────┐
    │ Constructed Code │
    │     Testing      │
    │ Executable system│
    └──────────────────┘
```

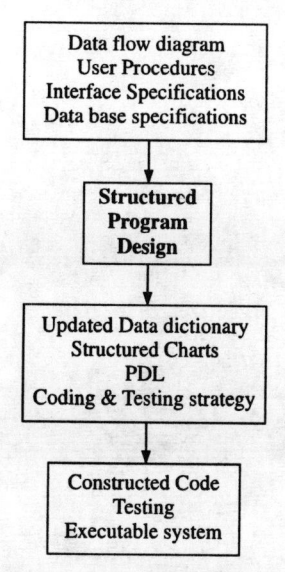

Figure 5.1 Program design.

Figure 5.1 illustrates the program design process and is summarized as the following:

- The inputs to the design process are the DFD, user procedures and interfaces and database specifications.
- The outputs of the design process are the updated design repository or data dictionary, package program specifications including structure charts and PDL for each module. This forms the input to constructing the application.
- The final output of the construction process is working program, testing strategy for the program.

5.2 STRUCTURE CHARTS

A structure chart is a graphic representation of program structure that depicts the modules of a program and the information that is exchanged between modules.

Shown in Figure 5.2(a) are the three symbols that are used to depict a structure chart:

1. Modules
2. Module connections
3. Information flows or data couples

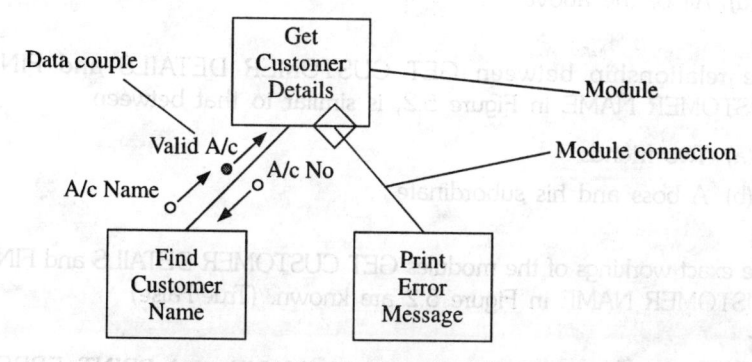

Figure 5.2(a) Structure chart symbols.

In the Figure 5.2(a), GET CUSTOMER DETAILS, FIND CUSTOMER NAME, PRINT ERROR MESSAGE are modules. Ac/No, A/c Name are data couples. If you compare the above structure chart, it can easily be transformed to either a paragraph perform or a section in your programming language.

The module GET CUSTOMER DETAILS invokes the module FIND CUSTOMER NAME. The module FIND CUSTOMER NAME receives the data couple A/c No, and if the A/c No is found, returns A/c Name along with Valid A/c flag.

Also notice, in the module PRINT ERROR MESSAGE, a small diamond connects it with GET CUSTOMER DETAILS module, which means it's a conditional invocation.

The symbol ↗ is known as **Control Couple**.

Difference between Data and Control Couples

There are two important differences between data and control couples.

1. Data couples are processed, whereas a control couple is not really processed.
2. A data couple may be manipulated by another data couple (e.g. multiplying quantity with unit price) or displayed on screen.

Exercises

#1
The symbol for a data couple is _____.
The symbol for a control couple is _____.

#2
The equivalent of a module in programming language is:

(a) PROGRAM
(b) PARAGRAPH
(c) Statement
(d) All of the above

#3
The relationship between GET CUSTOMER DETAILS and FIND CUSTOMER NAME in Figure 5.2, is similar to that between

(a) Two friends
(b) A boss and his subordinate

#4
The exact workings of the modules GET CUSTOMER DETAILS and FIND CUSTOMER NAME in Figure 5.2 are known. (True/False)

#5
Both the modules, FIND CUSTOMER NAME and PRINT ERROR MESSAGE, are always invoked by GET CUSTOMER DETAILS. (True/False)

#6
The decision to print the error message is made within PRINT ERROR MESSAGE.

Solutions to Exercises

1.

2. d
   The module can be any of the three options.
3. b
4. False
   However this lack of information about the working of modules does not hamper drawing the structure chart.
5. False
   PRINT ERROR MESSAGE is invoked only in case of an error, which is indicated by A/c valid flag.
6. False
   The boss module GET CUSTOMER DETAILS decides whether the error message should be printed, and appropriately invokes PRINT ERROR MESSAGE.

Iterative Invocation

Consider the following structure chart for accepting new customer details as shown in Figure 5.2(b). Observe the curved arrow below the module GET VALID A/C NO. This symbol represents **iteration**. It indicates that the three modules below GET VALID A/C NO will be repeatedly invoked until a specific condition is met.

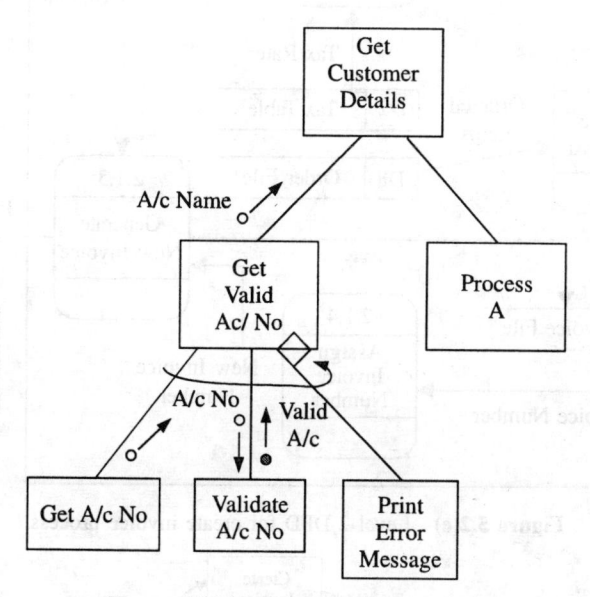

Figure 5.2(b) Iterative structure chart.

This is similar to "DO WHILE" loop in programming language.

Converting Level-2 DFD into Structure Charts

Figure 5.2(c) illustrates the level-2 DFD for Creating Invoice process. As per level-2 DFD, it illustrates all the sub processes and the data stores to and from which these processes will source their information. Next step in SSAD and the first of the steps in program design is to convert these data flow diagrams into structure charts.

Create Invoice structure chart in Figure 5.2(d) illustrates the importance of structured techniques: how an entire program, with dozens of lines of code can be arrived by converting DFD into structure charts. An analyst can build, evaluate and refine the structure chart until the design is so broken that it can be easily intercepted by the developer for working out the process specifications. Thus, when the designer is satisfied with the design in respect to complete representation of processes, the detailed program specifications for programming can be written using one of the techniques of *process specifications*, namely: *Structured English, Decision trees, Decision tables*.

92 Workbook on Systems Analysis and Design

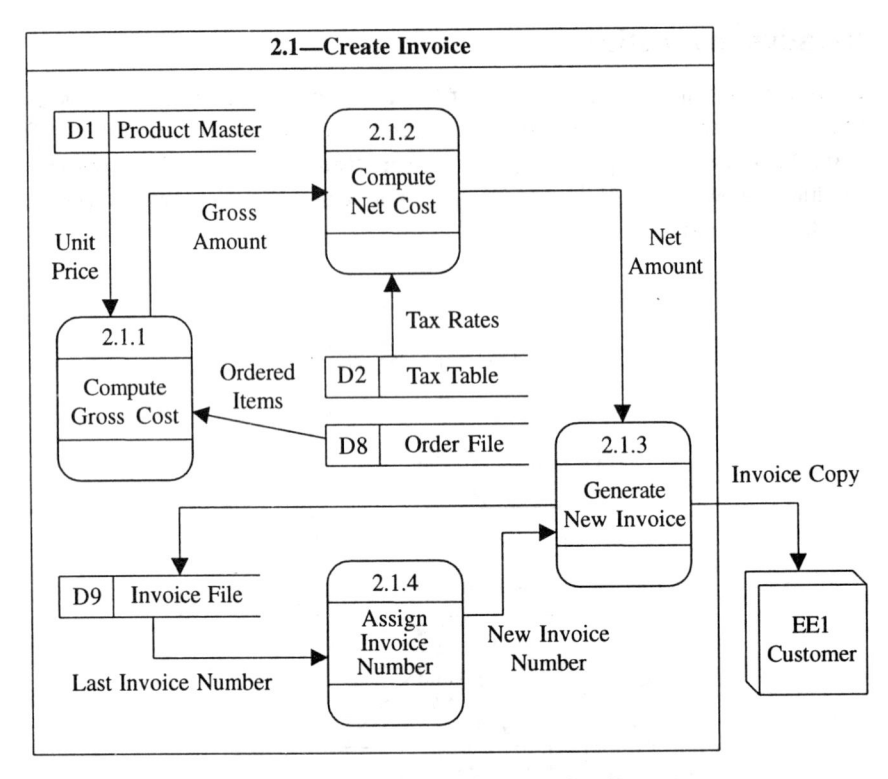

Figure 5.2(c) Level-2 DFD for create invoice process.

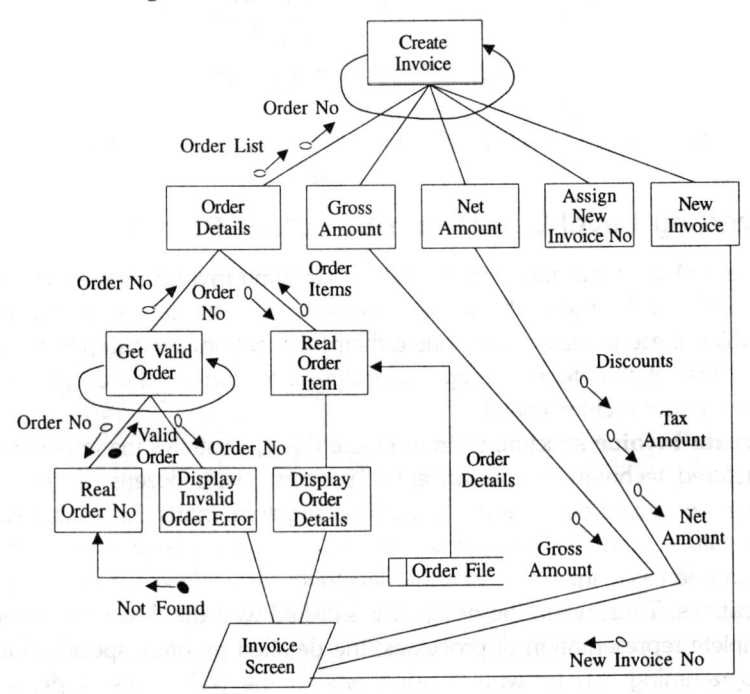

Figure 5.2(d) Completed structure chart for create invoice process.

Practice for Section 5.2

#1
Given below is a process specification for GET ORDER, by way of structured English.

Transform the same to a structure chart.

```
Get Order
For each part in order
    Check availability in inventory
        IF not available
            THEN Report Shortfall
                CASE reply
                    Cancel Order:
                        Report cancellation
                    Reduce Quantity:
                        Subtract from quantity ordered
                    Remove Parts:
                        Delete part from order
                END
            END
        END
    Store Order
```

#2
Given below is another process specification for GET VAILD ORDER, by way of structured English.

Transform the same to a structure chart.

```
Get Valid Order (Inputs are Valid-process-flag, Valid-invoice-flag, Order No,
Invoice Date)
SET valid-order-flag to false
SET valid-process-flag to true
REPEAT UNTIL no more Orders or Valid-process-flag is false
    CALL READ
        Read invoice file for input invoice detail (Order No & Invoice Date)
        IF valid-invoice-flag is false
        THEN
            CALL Validate Order No (Order No, Order-NOT-found)
            IF Order-NOT-found
                THEN CALL Write Invoice Error Screen (Order-NOT-found)
            END-IF
        END-IF
END REPEAT
END Get Valid Order
```

5.3 PROCESS SPECIFICATIONS

The structured techniques introduced earlier in previous chapters are excellent for documenting the procedures of existing and proposed systems. Data flow diagrams are especially useful as long they are simple enough for reader to understand.

Depending on the complexity of the procedures and the necessity of precise definitions the development team may need to use **Process specifications** to supplement structured and graphic techniques. The following three tools are used to specify the process logic.

While process specifications are normally used for lowest-level processes, it can be effectively used to prepare process descriptions that cannot be shown on data flow diagrams.

1. Process narratives
2. Decision Trees
3. Decision Tables
4. Structured English

Some process descriptions involve a number of different conditions that may occur in different combination, with each combination producing a specific outcome.

5.3.1 Process Narratives

Process narratives are, basically, verbal descriptions of process, which are brief. The narratives should be brief and as specific as possible. They may be used to describe special timing requirements, system constraints, or relationships among processes. The example shown below is how processes are described.

Example

Process Description

SYSTEM: TAB Checking Account 　　DATE PREPARED:
PROCESS ID: 1　　　　　　　　　　　PREPARED BY:
PROCESS NAME: Prepare Monthly Statement

PURPOSE: Prepare all checking account statements for a specified cycle

INPUT: TRAN-AMT
　　　　MONTHLY-STMT-TOTAL
　　　　NAME
　　　　ADDRESS
　　　　CURRENT-BAL

OUTPUTS: MONTHLY-STMT

> **PROCESS DEFINITION:**
> In order to balance workload, monthly statements are prepared in three cycles according to the following schedule:
> Cycle 1 = First business day on or after 1st of month
>   All personal accounts with odd sequence number
> Cycle 2 = First business day on or after 10th of month
>   All personal accounts with even sequence number
> Cycle 3 = First business day on or after 20th of month
>   All commercial accounts

5.3.2 Decision Trees

Decision Trees provide a graphic representation of decision logic that helps non-computer people find easy to understand.

How to construct decision trees?

The principles for development of decision trees are relatively straightforward.

➤ Identify all conditions
➤ Find out values these conditions may take or assume
➤ List all possible outcomes

> ***Example***
>
> The following represents a policy statement of customers doing business with the company. If the customer is doing business worth more than Rs. 1,00,000/-, he will get priority treatment by the company, whereas the customer doing business less than Rs. 1,00,000/- would get normal treatment. This can be illustrated by decision tree as shown below:
>
> ```
>                    Good Payment History ──────── High Priority
>        More than  ╱
>        Rs. 1 lac ╳
>        business   ╲                   With us more
>      ╱             Bad Payment ╱──── than 10 yrs. ──── High Priority
>     ╳              History    ╳
>      ╲                         ╲──── With us less
>        Less than                     than 10 yrs. ──── Normal
>        Rs. 1 lac                                       treatment
>        business ─────────────────────────────────── Normal
>                                                     treatment
> ```
>
> **Figure 5.3(a)** Decision trees.

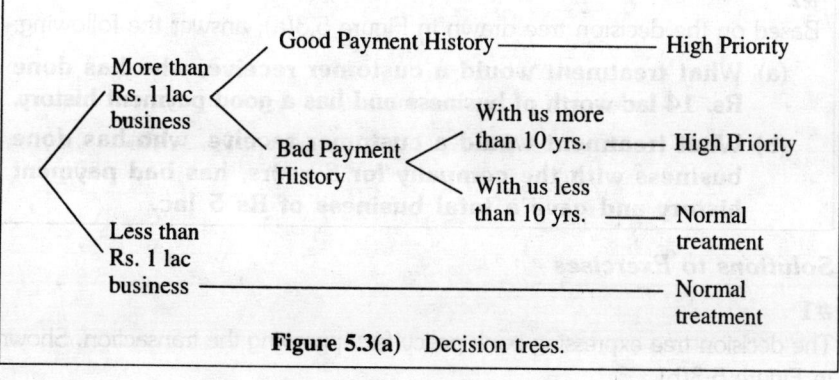

In the above example, there are three conditions which are effected: amount of business, payment history, duration of business. Based on the three conditions, each will take one of the values, based on which further branching will be done.

96 Workbook on Systems Analysis and Design

If we look further, the priority treatment is decided for customers doing business above Rs. 1 lac, depending on their last 10 years payment history. For each value (or set of values, like greater than Rs. 1 lac), a decision variable can take one or more paths. Subsequently this path would split into 2 or more paths, based on the set of values the second variable takes on, and so on.

Exercises

#1

Listed below are the three conditions along with the possible values and the three possible outcomes. The conditions and values will determine the outcome for approving of loans in bank. Draw a decision tree to illustrate the scenario.

Conditions and values

- CURRENT ACCOUNT BALANCE: values >= 1000 or < 1000
- NUMBER OF OVERDRAFTS: values <= 2 or > 2
- AVERAGE SAVINGS BALANCE: values >= 500 or < 500

Possible outcomes for approving the automatic loan on overdrafts.

- Approval (no limit): For current balances greater than or equal to 1000 and number of overdrafts less than or equal to 2
- Conditional approval (Rs. 500): For current balances greater than or equal to 1000 and number of overdrafts more than 2 but average balance being greater than or equal to 500. Also for current balance less than 1000 and number of overdrafts not more than 2 and average balance being not less than 500.
- For all other conditions: Rejection of transaction

#2

Based on the decision tree drawn in Figure 5.3(a), answer the following:

(a) **What treatment would a customer receive, who has done Rs. 14 lac worth of business and has a good payment history.**

(b) **What treatment would a customer receive, who has done business with the company for 8 years, has bad payment history and gave a total business of Rs 5 lac.**

Solutions to Exercises

#1

The decision tree expressing bank policy for approving the transaction. Shown in Figure 5.3(b).

#2

(a) High priority

(b) Normal treatment

Program Design **97**

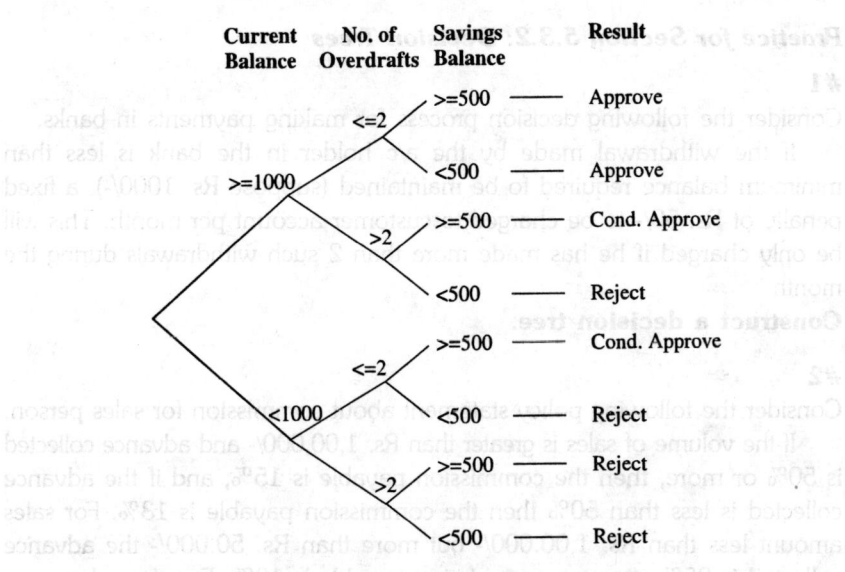

```
Current    No. of    Savings     Result
Balance  Overdrafts  Balance
                       >=500  ——— Approve
                <=2
                       <500   ——— Approve
       >=1000
                       >=500  ——— Cond. Approve
                >2
                       <500   ——— Reject

                       >=500  ——— Cond. Approve
                <=2
        <1000          <500   ——— Reject

                       >=500  ——— Reject
                >2
                       <500   ——— Reject
```

Figure 5.3(b) Decision tree for Exercise 1.

Practice for Section 5.3.2: Decision Trees

#1
Consider the following decision process for making payments in banks.

If the withdrawal made by the a/c holder in the bank is less than minimum balance required to be maintained (suppose Rs. 1000/-), a fixed penalty of Rs. 50/- to be charged to customer account per month. This will be only charged if he has made more than 2 such withdrawals during the month.

Construct a decision tree.

#2
Consider the following policy statement about commission for sales person.

If the volume of sales is greater than Rs. 1,00,000/- and advance collected is 50% or more, then the commission payable is 15%, and if the advance collected is less than 50% then the commission payable is 13%. For sales amount less than Rs. 1,00,000/- but more than Rs. 50,000/- the advance collected is 25% or more, commission payable is 10%. For any sales more than Rs. 50,000/- the advance collected is less than 25%; then the commission payable is 8%. For sales other than mentioned above no commission is payable.

List the different conditions and the different values it will take. Also draw a decision tree for the same.

5.3.3 Decision Tables

Using decision tables, decision trees' conditions and outcomes are listed in the form of a two-dimensional table. A decision table, as compared to a decision tree, checks all the possible combinations that might arise for all conditions.

General format of decision tables

List of conditions	Columns representing logical combinations of conditional values
List of outcomes	Resulting outcome for each set of conditions

The left top column lists condition, and the right hand side (r.h.s.) column gives the corresponding procedures for each possible condition.

> **Example**
>
> Consider the Exercise no 1 on p. 90 in decision trees for: approving the bank loan based on current balance, number of overdrafts and average savings balance.

The following decision table may be drawn for these conditions.								
Current balance >= 1000	Y	Y	Y	Y	N	N	N	N
Number of overdrafts <= 2	Y	Y	N	N	Y	Y	N	N
Average savings balance >= 500	Y	N	Y	N	Y	N	Y	N
Approve	X	X						
Condition approve			X		X			
Reject				X		X	X	X

Figure 5.3(c) Decision table.

To fill in the decision table, first determine the number of rows and columns. Then take the last row in the condition stub. Fill in the row with the pattern of different possibilities that the condition might take.

Example

Consider the following decision table shown in Figure 5.3(d), which expresses a policy statement.

	Rules							
	1	2	3	4	5	6	7	8
Conditions								
C1: more than 10000 business	Y	Y	Y	Y	N	N	N	N
C2: good payment history	Y	Y	N	N	Y	Y	N	N
C3: with us for more than 20 years	Y	N	Y	N	Y	N	Y	N
Action								
A1: Type of treatment P-Priority, N-Normal	P	P	P	N	N	N	N	N

Figure 5.3(d)

Each column of the condition column would give a rule and each column of the action, would give the action/decision/conclusion for this rule. Like Rule 3 states that if the customer has given business greater than 10000, has a bad payment history and has been with us for more than 20 years then he gets priority treatment.

Exercises

Based on the above Example and Figure 5.3(d), answer the following.

#1
What does the rule 6 state?

#2
What type of treatment does a customer get if he has a good payment

history not doing business more than 10000 and with us for more than 20 years?

#3

The conditions and the various values each condition can take, are used to draw a decision table, and is same as those for a decision tree (TRUE/ FALSE).

Solutions to Exercises

#1

The customer has given business less than or equal to 10000, has good payment history but has been with us for less than or equal to 20 years; then he would get normal treatment.

#2

Normal

#3

True, it is exactly same as a decision tree.

Practice for Section 5.3.3: Decision Tables

#1
"If the volume of sales is greater than Rs. 10000 and advance collected is 50% or more than the commission is 16%. If the advance collected is less than 50% then it is 14%. For sales Rs. 10000 irrespective of the advance collected, commission is 10%. For sales less than 10000 commission is 9% or 8% based on whether advance collected is 50% more or less respectively.

Based on the narrative:

(a) List the conditions
(b) List the values associated with each condition
(c) Develop a decision table based on the decision tree.

#2
A minimum of 5% discount applies for all purchases. If the retailer maintains an average monthly purchase volume of at least 100,000, a 15% discount applies, provided the retailer is an AGA member. When the retailer's purchase volume is less than 100,000 the discount is 12% for AGA members and 9% for non-members. Retailers who are non-members but who maintain an average of 100,000 monthly purchase, qualify for 10%, unless the purchase totals is not less than 40,000 in which case the 5% general discount is applicable.

Based on the narrative:

(a) List the outcomes
(b) List the conditions
(c) List the values associated with each condition
(d) Develop a decision tree relating these conditions, values and outcomes.
(e) Develop a decision table based on the decision tree.

#3
What do decision trees and decision tables have in common? How do they differ?

#4
A firm pursues the following discount policy on its products (all discounts being offered as a percentage of advertised prices).

Those customers ordering more than 10 items receive at least 2% discount. Those customers who are not regular customers receive at most 3% discount. All those regular customers who order more than 10 items receive a 3% discount plus an additional 2% if they pay in cash. Any customers paying cash and who orders more than 10 items receive 4% discount. Any customer not covered by the preceding rule, receives no discount.

By using decision trees and decision tables, evaluate the above rules for their consistency, comprehensiveness and redundancy (if any).

#5

An international airline initiated a frequent-flier program to encourage its passengers to fly regularly. The policy of this program is:

"Passengers who have logged more than 200,000 km of flying time during the current *financial year* are to receive a free trip to Singapore and Hong Kong. Similarly passengers who have flied more than 100,000 km per *calendar year* and, in addition, paid cash for tickets or have been flying the airline regularly for more than 2 years, would also receive the free trip to Singapore and Hong Kong. Passengers who fly less than 100,000 km but more than 75,000 km per calendar year, but had been regularly flying with the airline for the last 3 years, are also entitled for the free trip"

Based on the narrative:

(a) **List the outcomes.**

(b) **List the conditions.**

(c) **List the values associated with each condition.**

(d) **Develop a decision tree relating these conditions, values and outcomes.**

(e) **Develop a decision table based on the decision tree.**

5.3.4 Structured English

As verbal statements are a natural medium of communication between users and programmers/analyst, a series of formal structured English statements can be used effectively to communicate processing rules. Structured English can be used to specify any process logic. Decision trees and decision tables are used to specify only decision logic.

Structured English (referred also as Pseudo code) as a specification language makes use of limited vocabulary and a limited syntax. The vocabulary mostly consists of:

➢ Imperative English language verbs

➢ Terms defined in the data dictionary

➢ Certain key words for logic formulation

How to illustrate?

There are three important things a pseudo code illustrates:

Sequence. Stringing statements together one after the other. They can be simple commands or processing blocks arising from one of the other constructs.

Selection. Using IF-THEN-ELSE one can communicate possible selection of process or set of activities.

Example
IF customer does more than X business
    And IF (customer has good payment history)
        THEN priority treatment
    ELSE (bad payment history)
        IF customer doing business for more than Y years
            THEN priority treatment
        ELSE
            Normal treatment
Else
    Normal treatment

Iteration. Used to represent a sequence of steps that are to be repeated specified number of times.

Example
For each day of the month
    Set DAILY-INT to DAILY-BAL times DAILY-RATE
    Add DAILY-INT to MONTHLY-INT
Add MONTHLY-INT to YEAR-2-DATE

When translated into flow charts, these look as shown in Figure 5.3(e).

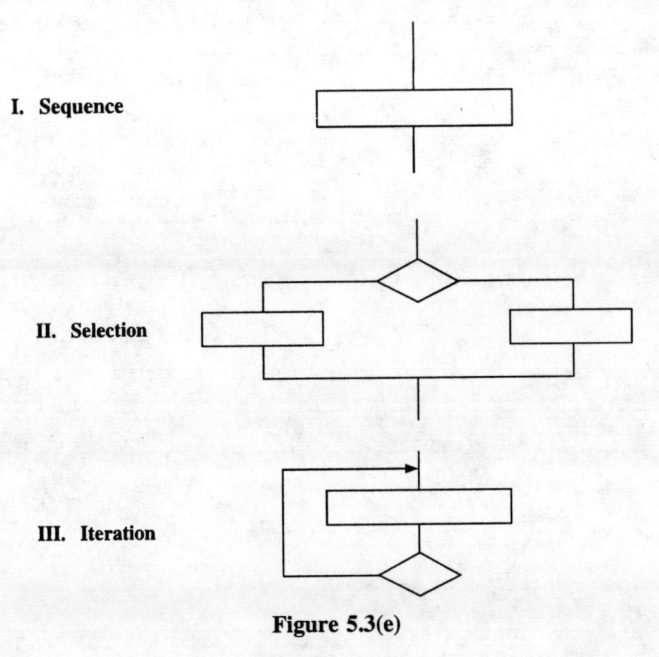

I. Sequence

II. Selection

III. Iteration

Figure 5.3(e)

Exercise

Write the following specification in structured English.

If the volume of sales is greater than Rs. 10000 and advance collected is 50% or more, then the commission is 16%. If the advance collected is less than 50% then it is 14%. For sales equal to Rs. 10000 irrespective of the advance collected, commission is 10%. For sales less than Rs. 10000 commission is 9% or 8% based on whether advance collected is 50% or more, or less than 50% respectively.

Solution to Exercise

IF sales greater than 10000
And IF advance collected greater than or equal to 50%
        THEN commission is 16%
    ELSE (advance less than 50%)
        Commission is 14%
ELSE (sales not greater than 10000)
    IF sales equal to 10000
        THEN commission is 10%
    ELSE (sales less than 10000)
        IF advance collected greater than or equal to 50%
            THEN commission is 9%
        ELSE
            Commission is 8%

Practice for Section 5.3.4: Structured English

#1
Given below is a process specification as narrated by a user.
Convert the specification into Structured English.
  Policy for Deletion of suspended Accounts. We go through the account file, one customer record at a time. For each customer record we do the following. First, we ask if it has been marked "Suspended" in the account-status indicator. If so, we ask if the date of last transaction is more than 30 days before today's date. If that is also true, we do the following four things:

- Mark the account status "Retired".
- Set the date of last transaction to today's date.
- Write a copy of the customer record to the history log.
- Delete the customer record from the account file.

#2
A stack of student records, ordered by student no., contains up to five records on each student covering the student's exam results in one or more of the following—accounting, economics, law, mathematics, computers. A clerk is required to process these records and must produce an average for each student, which is then entered on a summary sheet. Not all students take all five subjects. The clerk also computes an average mark for each exam for each of the five subjects and enters it on the summary sheet.
Produce a structured English specification to translate into process specifications.

#3
System analyst has provided the following written description:
  The number of days' leave to which an employee is entitled depends upon how long he has worked with the company at the start of the holiday year. The leave is counted for a calendar year from 1st Jan to 31st Dec. If he has worked less than one year, then he gets one and two-third days for every complete month. If he started after 30th October in the previous year, then he get no leave this year. If he has worked more than one year, then he gets 20 days leave. If he has worked for more than 5 years, then he gets an extra two days, and if he has worked for more than 10 years, he gets additional 3 days.
  Indicate how this could be represented in:

(a) **Decision table**
(b) **Sequence of structured English.**

#4
Write structured English description corresponding to the following narrative:
  The sales clerk totals the prices of the items on the order and enters

this on the customer's account as a debit. She then checks the customer's name and address card to see where the goods are to be delivered. If the client lives more than 50 miles from the depot, then a delivery charge of 2% is added to the order. The delivery charge is then debited to the customer's account. But if the customer's business over the last three months is more than five times the value of the current order, then no delivery charge is made. If a delivery charge is made, then it is credited to the general carriers account.

#5

Shown below a code written down by means of pseudo code. Prepare a decision tree for the same.

```
DO WHILE MORE EMPLOYEES
    IF EMPLOYEE NOT EXITS
    ELSE
        IF GRADE <=20
            DA = 15% OF BASIC
            HRA=300
        ELSE
        IF GRADE <= 35
            DA = 20% OF BASIC
            HRA=600
        ELSE
        IF GRADE < 40
            DA = 30% OF BASIC
            HRA=800
        ELSE
            DA = 40% OF BASIC
            HRA = 20% OF BASIC
    ENDIF
ENDDO
```

5.4 PROGRAM CODING STANDARDS

Constructing the System Components

Now that the design has been completed and approved, you are ready to begin coding the system. Constructing a system is similar to constructing a house. Both activities begin with the foundation. In system construction, the facility being used is the foundation. The facility will include the tools, hardware, software, programming languages, database engine, etc.

But having the sophisticated tools and using them will not deliver a good system. A lot depends on the standards followed during the construction of the programs.

Conventions for Coding Structured Programming

A primary goal of structured programming is to create code that is as clear and readable as possible. A program that is clear and readable is easier to develop, test, debug, and maintain than one that is not. Programming standards are guidelines adopted by an organization's programming staff so as to produce maintainable and flexible systems. These standards specify conventions for naming data elements and documenting program code.

Here are some guidelines that will help you create more readable code.

Use only proper structures

The basic theory of structured programming is that any program can be written using three logical structures:

- Sequence
- Iteration
- Selection

As a result, a program made up of these structures will have only one entry point and one exit point. Thus the program is executed in a controlled manner from the first statement to the last.

Sequence. The sequence structure, is simply a set of imperative statements executed in sequence, one after another. The exit point is after the last function in the sequence. A sequence may consist of a single or many functions. Simple statements like PERFORMs or CALLs are the prime example for sequence structure.

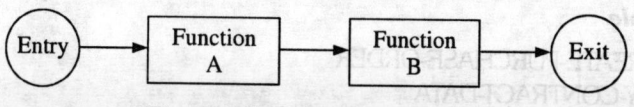

Figure 5.4(a) Sequence structure.

Selection. The selection structure, is a choice between two, and only two, functions based on a condition. This structure is often referred to as the IF-THEN-ELSE structure.

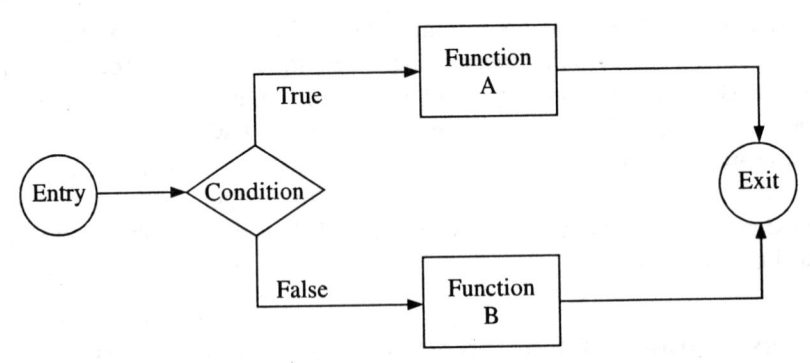

Figure 5.4(b) Selection structure.

Iteration. Iteration is the continued activation of a module or a set of statements for as long as a stipulated condition exits. When the condition is no longer true, the program continues with the next structure. DO-WHILE or DO-UNTIL and PERFORM-UNTIL are some of such functions.

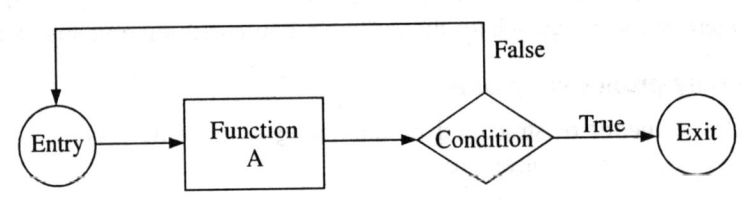

Figure 5.4(c) Iteration structure.

Use module that relate to the program structure chart

Two **benefits** result when you create module/paragraph names which denote the function associated with it.

1. First, the program structure chart becomes a directory to all the modules of the resulting programs, by which any future problem modules can be located by studying the structure chart.
2. Secondly, the search for locating these modules becomes easy in programs, which has so many of such functions and runs into hundreds of lines of codes.

Example

000-CREATE-PURCHASE-ORDER
VERIFY-CONTRACT-DATA

Use descriptive data names

When you create data names, do not abbreviate unnecessarily. Make the names as descriptive as you can within the constraints of your programming language.

Also use prefix/suffix to data names to distinguish from one data structure to another.

Example
AC-OPEN-DATE
TRAN_AC_CODE
TRAN_AMOUNT

Use comments sparingly

The structure and code of a program should make it easy to understand without comment. However, comments can be useful to overcome language limitations, or to provide a reference to documentation that is external to the program.

Conclusion

Each application software program should be designed to accomplish as much as possible within defined specifications. Always remember that a program is an implementation of a plan to solve a business problem. Hence never concentrate on elegant program designs. Instead concentrate on program efficiencies and refinements that can be easily implemented.

5.5 SOFTWARE TESTING TECHNIQUES

One of the Murphy's laws states that anything that can go wrong will go wrong. This is especially true in system's development. Hence the system must be thoroughly tested to reveal its errors and misconceived functions before it is finally delivered to the user.

Planning the testing of a programming system involves formulating a set of test cases, which are similar to the real data that the system is intended to manipulate. Test cases consist of an input specification, a description of the system functions exercised by that input and a statement of the extended output. Thorough testing involves producing cases to ensure that the program responds, as expected, to both valid and invalid inputs, that the program performs to specification and that it does not corrupt other programs or data in the system.

In principle, testing of a program must be extensive. Every statement in the program should be exercised and every possible path combination through the program should be executed at least once. In practice, this is impossible in a program, which contains loops, as the number of possible path combinations is astronomical.

The best that can be done in reality is to derive test cases, which cause each path through the program to be executed. This also ensures that each statement in the program is executed at least once. Thus, it is necessary to select a subset of the possible test cases and conjecture that this subset will adequately test the program.

Approach to Testing

- Testing a system's capabilities is more important than testing its components. This means that test cases should be chosen to identify aspects of the system, which will stop them doing their job.
- Testing old capabilities is more important than testing new capabilities. If a program is a revision of an existing system, users expect existing features to keep working.
- Testing typical situations is more important than testing boundary value cases. It is more important that a system works under normal usage conditions than under occasional conditions, which only arise with extreme data values, like volume, size, etc.

Test Cases and Test Data

Test cases and test data are not the same thing. Test data are the inputs, which have been devised to test the system, whereas test cases are input and output specifications plus a statement of the function under test. It is sometimes possible to generate test data automatically, but impossible to generate test cases.

Creating a Test Plan

Any test plan should cover the following:

1. Test groups. This is to ensure that testing as an activity is carried out in an organized manner. The test group is formed who will be responsible for developing the test data and for carrying out tests at each stage.

2. Preliminary tests of each component. During preliminary test, the components are tested individually to determine their ability to meet acceptance criteria, i.e. the functional requirements for the program are met.

3. Final tests. After all the individual components have been tested and shown to meet the acceptance criteria, they are brought together and tested as a unit.

4. Documentation of test plans, data and results. Since testing is a planned process that is critical to the success of the system, it requires extensive documentation. The documentation should detail the entire test plan, including the specific tests to be conducted, the sequence in which they will be conducted, and the tentative test schedule.

Figure 5.5(a) shows a "Test Specification form". This is the standard documentation form used to create test plans, test cases and test results.

In the Figure 5.5(a), the user has created a test specification for a data entry program. He has listed some of the conditions which will be tested at predefined steps, and also the expected error results.

Test Specification			Page 1	
System: Medicine Prescription system			Module: Data entry	
Plan Prepared by: Reshmi			Date 1/10/97	
Objective: To test data validation and error reporting				
Condition	Description	Test Steps (all messages)	Expected Results	Tested By
011	Invalid doctor code	Enter registered doctor no	"Doctor not registered"	
012	Invalid charge code	Enter charge code "G" "W" "D"	"Invalid charge code, G/W/D"	
013	Drug code not valid	Validate as per check list	Drug code not defined	
014	Drug not matching for illness	Illness versus Drug code	Drug prescribed not for illness	

Figure 5.5(a) Examples of test data as per specified format.

Testing Techniques

Testing can be formalized onto 4-phase process, as follows:

1. Unit testing: A single program. Unit testing is the basic level of testing where individual components (which may be functions/routines/programs) are tested to ensure that they operate correctly. In a properly designed system, each component should have a precise specification, and test cases must be defined to check that the component meets its specification.

Unit testing considers each component to be a stand-alone entity, which does not require other system components to be present during the testing process.

2. Integration testing: A related group of program collectively known as modules. A module is a collection of programs, which are interdependent. After each program unit has been tested, the interaction of these programs when they are put together must be tested. A module encapsulates related functions and routines and it should be possible to test a module as a stand-alone entity, without the presence of other system modules.

3. System Function testing: The entire software application module. System testing is the next step up in the testing process where different modules are put together to form subsystems. Each of the modules may be designed and implemented by different software engineers, hence there may be problem with respect to interface mismatches. Thus, the system test process should concentrate on the detection of the interfacing errors by rigorously exercising these interfaces.

4. Acceptance testing: The entire software program with live data for production readiness. Acceptance testing is the process of testing the system with real data – the information, which the system is intended to manipulate. Acceptance testing often demonstrates errors in the system requirement's definition. The requirements may not reflect the actual facilities and performance required by the user and testing may demonstrate that the system does not exhibit the anticipated performance and functionality.

Ideally these phases should be completed sequentially. However, when development work is done in parallel, module coding, unit testing, integration testing is commonly integrated, followed by system testing and acceptance testing.
Figure 5.5(b) illustrates the complete cycle of software/application testing.

Choosing an Approach to Testing

There are two approaches to testing accuracy; black box and white box testing. Although both are designed to create accurate systems, one focuses on error-free functions and the other on error-free code.

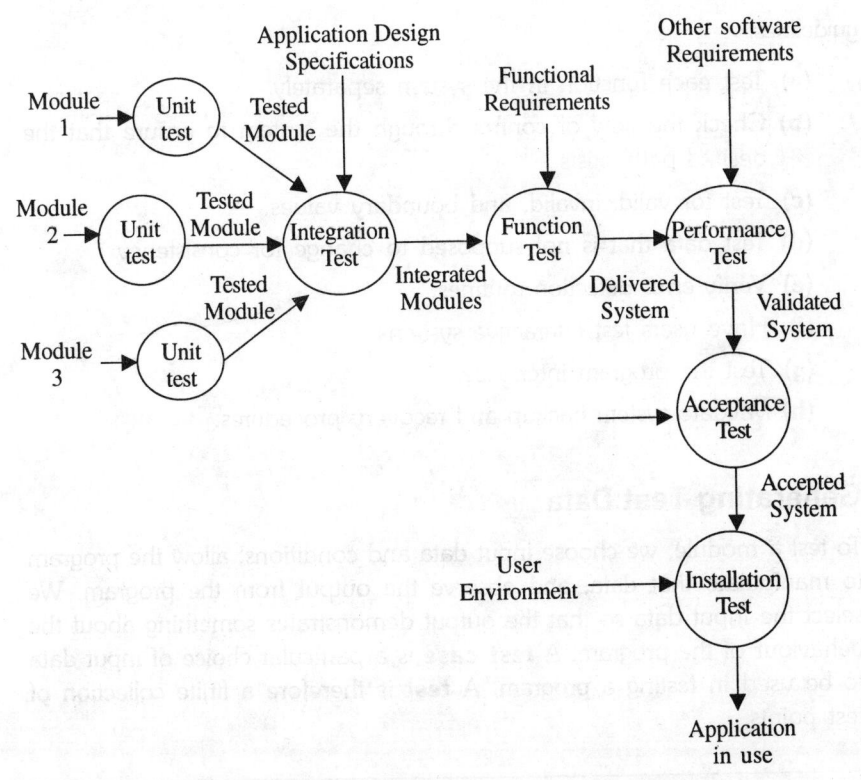

Figure 5.5(b) Complete testing stages in testing cycle.

Black Box testing is the name given to testing where the tester is presented with the specification of a component to be tested and uses this to derive the test cases. This approach has the advantage that testers need not have access to the source code of the routine and need not understand the program, which is being tested. This disadvantage, of course, is that the tester cannot get clues from the program about which test inputs best exercise the program.

An alternative approach to testing is called **white box or structural testing** in that it relies on the tester using knowledge about the code and the structure of a component to derive the test data. The advantage of structural testing is that test cases can be derived systematically and test coverage measured. Thus the quality assurance mechanisms which are set up to control testing can quantify what level of testing is required and what has been carried out.

Building Test Data Files

In order to test the running of a system, you need data. Test Data should force the system to traverse every possible path. Here are some useful guidelines:

(a) Test each function in the system separately.
(b) Check the flow of control through the system to ensure that the desired path exists.
(c) Test for valid, invalid, and boundary values.
(d) Test data that is not supposed to change for consistency.
(e) Verify error detection routines.
(f) Have users test interactive systems.
(g) Test the program interfaces.
(h) Validate system backup and recovery procedures.

Generating Test Data

To test a module, we choose input data and conditions, allow the program to manipulate that data, and observe the output from the program. We select the input data so that the output demonstrates something about the behaviour of the program. A **test case** is a particular choice of input data to be used in testing a program. A **test** is therefore a finite collection of test points.

> *Example*
> Consider a program that, given the coefficients of the variables, calculates the two roots of a quadratic equation. If our test objective is to demonstrate that the program functions properly, we might choose test points where the coefficients A, B, and C range through representative combinations of negative numbers, positive numbers, and zero. For example, we may choose combinations so that
>
> 1. A is greater than B and C
> 2. B is greater than C and A
> 3. C is greater than B and A
>
> For instance, if the module is coded to expect an input variable that is supposed to be positive, a test case may be included for each of the following:
>
> 1. A very large positive integer
> 2. A positive integer
> 3. A positive fixed point decimal
> 4. A number between 0 and 1
> 5. A negative number or 0 itself

Testing the Hardware Component

Often systems are developed using an organization's existing hardware resources. In such situations, you will need to test the hardware on which the applications will run, to make sure that it can handle the added load your application will place on it.

Some of the common hardware tests are:

1. Test the access speed	Are there enough communication channels to ensure an acceptable response time when a user logs onto a terminal while running the application.
2. Test the processing speed	Does the hardware meet the throughput or turnaround requirements of the users' production schedule – does it process transactions at an acceptable rate?
3. Test the storage capacity	Do the disk drives store the volume of data required?
4. Test the peak load	When all the users are logged into the applications, doing all possible operations, Data entry, Processing, Printing, Backups, etc, does turnaround time, access speed or response time decrease to unacceptable level.

Software Testing Cycle

Software error means that the software does not do what the process specification documents describe.

Example
The requirement specification may state that the application must respond to a particular query only when the user is authorized to see the data. If the program responds to an unauthorized user, then that application is not working properly.

There are several reasons for the system to give erroneous results, the major being:

- The process specifications may be wrong. It may not be what the customer really wants or needs.
- The specification may specify something that may not be possible for the prescribed Hardware and software on which the application begins.

- The system design itself may be at fault. The fault may be due to incorrect process links or database design.
- The program design may be at fault. The algorithm for the process may not be laid down correctly.
- The program code itself may be program. The code may implement the access algorithm incompletely.

By the time we are testing the program modules, the developer hopes that the specifications are correct. Moreover, having used the software engineering techniques, we have tried to assure that the design of the system and program modules reflects the requirements and forms a basis for a sound implementation. However, the stages of software development cycle involve not only computing skills but also our communication and interpersonal skills.

We say that a software system contains an error if it does not do what the customer expects it to do. A failure is an occurrence of an error somewhere in the software system. Thus, an error reflects the developer's view of the software, whereas a failure reflects a customer's view. The customer reports a failure, and the developers must find and correct the errors that resulted in that failure.

Types of Errors

After coding the program modules, we usually examine the code to spot errors and to eliminate them right away. When no obvious errors exist, we then test our program to see if we can isolate more errors. Thus, it is important that we know the kinds of errors for which to look. We can catalog the types of errors that occur in the coding of the program modules.

Algorithmic error

An algorithmic error is one in which a program module's algorithm or logic does not produce the proper output for a given input because something is wrong with the processing steps. These errors are sometimes easy to spot just by reading through the program. Typical algorithmic errors include:

1. Branching too soon
2. Branching too late
3. Testing for the wrong condition
4. Forgetting to initialize variables or set loop invariant
5. Forgetting to test for a particular condition (e.g. when division by zero might occur)
6. Comparing variables of inappropriate data types

Syntax errors

When checking for algorithmic errors, we may also check for syntax errors. Here, we want to be sure that we have properly used the constructs of the programming language.

Computation and precision errors

They occur when the implementation of a formula is wrong or does not compute the result to the required degree of accuracy. For example, combining integer and fixed- or floating-point variables in an expression may produce unexpected results. Sometimes, improper use of floating-point data or ordering of operations may result in less than acceptable precision

Documentation errors

When the documentation describes the program's function but does not match what the program actually does, we say that the program has documentation error. Often the documentation is derived from the program design and provides a very clear description of what the programmer would like the program to do, but the implementation of those functions is in error.

Capacity or boundary errors

They occur when the performance of a system becomes unacceptable as the activity on the system reaches its specified limit. For instance, if the requirements specify that a system must handle thirty-two devices, the programs must be tested to monitor the performance of the system when all thirty-two devices are active. Moreover, the system should also be tested to see what happens when thirty-three devices are active, if such activity is possible. By testing and documenting the system's reaction to a configuration beyond its capacity, the test team helps the maintenance team understand the implications of increasing system capacity in the future.

Throughput or performance errors

They occur when the system does not perform at the speed prescribed by the requirements. These are timing errors of different sort: time constraints are placed on the system's performance by the customer's requirements, rather than by the need for coordination.

Recovery errors

They can occur when an error is encountered and the system does not behave as the designers desire or as the customer requires. For example, if a power failure occurs during system processing, the system should recover in an acceptable manner. For many systems, a set of hardware and system software is prescribed in the requirements. The software modules are designed according to specifications in the hardware and system software documentation.

Hardware and system software errors

They can arise when the documentation for the supplied hardware and software does not match their actual operating conditions and procedures.

Standards and procedures errors

They may not affect the running of the programs, but they may foster an environment for the creation of errors as the system is tested and modified. By failing to follow the prescribed standards, one programmer may make it difficult for another to understand the programming logic or to find the data descriptions needed to solve a problem.

Practice for Section 5.5: Software Testing

#1
Assume that the Add student module in Student Registration system requires that the user enter the following data:

Student No.:	xxxxx (Numeric)
Student Name:	First name (max 15 characters), last name (max 15 characters)
Course code:	xxxxx (Alphabetic)
Course Fees:	xxxxxx (Numeric)
Start Date:	dd/mm/yy

(a) **Identify at least five data entry errors that might arise as this data is entered.**

(b) **Create a small set of test data to test for these error conditions.**

#2
Shown below in Figure 5.5(c), the Order Entry Screen. Assume for all the data items the necessary data types.

(a) **Draw out a plan for testing the necessary programs associated, before you test the Order entry.**

(b) **Identify at least five data entry errors that might arise as this data is entered.**

(c) **Create a small set of test data to test for these error conditions.**

```
Run Date xx/xx/xx        ORDER ENTRY SCREEN        Screen:1/2

ORDER NUMBER   [      ]      ORDER DATE (mm/dd/yy)  [ ][ ][ ]
CUSTOMER NUMBER [      ]     CUSTOMER NAME
ORDER STATUS CODE [      ]   [                              ]

ORDER ITEMS:
    ITEM NO     PRODUCT NO    QTY ORDERED    UNIT PRICE
       1        [       ]     [         ]    [        ]
       2        [       ]     [         ]    [        ]
       3        [       ]     [         ]    [        ]

Press F1 for Help of Customer Code
Press F2 for Product Number
Press F5 for Next Screen
Press F10 to exit

   MESSAGE:   [ Product No not defined in Product Master — Enter to continue ]
```

Figure 5.5(c) Order Entry Screen.

#3

Shown below in Figure 5.5(d), is Invoice Entry Screen. Assume for all the data items the necessary data types.

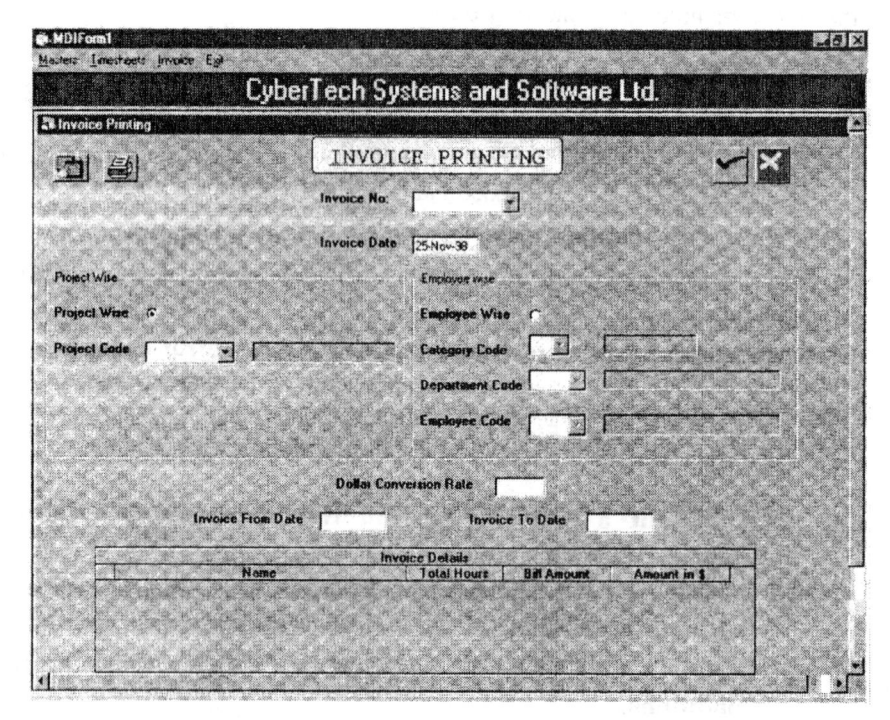

Figure 5.5(d) Invoice printing screen.

(a) Identify at least five data entry errors that might arise as this data is entered.
(b) Create a small set of test data to test for these error conditions.
(c) How would you document the tests performed and the results generated?

#4

Suppose a program contains N decision points, each of which has two branches. How many test cases are needed to perform path testing on such a program? If there are M choices at each decision point, how many test cases are needed for path testing?

6

Case Study

6.1 OVERVIEW

Anirudh Sharma runs a record shop by name *"Classic Collections"*. On an average, 1000 customers visit the shop daily, totaling sales in range of 500-600. The store has a collection of 2500 titles of cassette tapes, CDs both video and audio. For the most popular records, the store keeps as many as 250 copies per title.

Customers spend a lot of time browsing around the shelves or asking the counter staff for particular recordings. The store has also drive-in points, where one can listen to select hottest titles, before deciding on the purchase. There is a pickup desk at the shop wherefrom the customers can pick up a selection list (Figure 6.1(a)), and enter the titles they want to buy. The titles are identified by unique Number stick to the cover. This is then produced at sales counter for invoicing and delivery.

The current manual system, which the database is to replace, works in the following manner. When a sale is made, customers are given a receipt (Figure 6.1(b)) with details of the record they have bought. Jackie retains a carbon copy of this receipt and enters details from it on ledger (Figure 6.1(c)). The receipt is also used to update the stock record cards (Figure 6.1(d)). The volume of sales is quite considerable and has led to Classic Collections spending an increasing amount of time updating the stock record cards and ledger at the end of the day. Receipts are then filed in a loose-leaf folder.

If the customer cannot find the recording of their choice, they can place special orders. A form is available (Figure 6.1(e)), and customers are required to pay Rs. 25 as deposits for Cassette and Rs. 100 for CDs. When the order is received, the customer is informed through their contact no (if they have one), or through post cards to their address, to collect the orders (Figure 6.1(f)). The deposits are kept separate with details of the order and postcards have to be completed, addressed and paid for! Since this earns a good will for Classic Collections, they want to continue with this service.

If the customer do not respond to their orders within 7 days, the store can sell it to other customers. After the expiry of wait period, if the customer turns up and the disc still remains in the stock, he is asked to pay additional Rs. 10 for cassette and Rs. 20 for CD. In case the customer wants the refund, Rs. 10 is deducted for a cassette and Rs. 40, against the deposit amount. The store wants to vary the deposits, based on the price of the cassette or CD. The store wants to incorporate this facility in the new system.

An important aspect of the music store business is the ability to match supply and demand. Her stock should be sufficient to service requests, but should not be so large that it unnecessary ties up her limited capital. The store buys the cassettes and CDs from different wholesalers. The store decides the requirements every Monday while the shop is closed for sales and places a purchase order with each supplier (Figure 6.1(g)). The store keeps a lead-time (the time between records being ordered and delivered) of one week. Deliveries are received once or twice a week from each supplier accompanied by delivery note (Figure 6.1(h)). All goods received are checked for damage, and poor quality goods are returned with a covering note (Figure 6.1(i)).

The special orders are also included as part of normal orders to suppliers. Anirudh has appointed Cyber Systems to study, analyze, design and develop a computerized system to handle the above activities for his store. Imagine that you have been designated as system analyst by Cyber Systems to deliver the same.

Anirudh wants to expand the system to incorporate additional information. He wants the Disc to be identified if it is an audio/video cassette, or, audio/video CD. Also in the new automated system, if the sale desk finds any particular title to be selling hot or feels that more of a particular title to be ordered, it will mark those titles to be ordered.

Shown in Figure 6.1(a) through Figure 6.1(i), are the current manual records/documents maintained by Classic Collections, as part of their day-to-day operations at the shop.

Format of Customer Selection List (Figure 6.1(a))

Classic Collection
3, Krishna Towers
S.P. Mukherjee Road
Fort, Bombay 400 001

Selection

Code	Title and Artist	Type	Nos
WC107	50s Classics—Frank Sinatra	CD	1
HM545	Doli Saja Ke Rakhna	Cassette	1

One Stop Store of Music

Figure 6.1(a)

Customer Receipt on Sale (Figure 6.1(b))

```
Classic Collection
3, Krishna Towers
S.P. Mukherjee Road
Fort, Bombay 400 001

         SALE RECEIPT: 1737        Date: 1/1/97
```

Title	Unit Price	Qty	Amount
Kuch Kuch Hota Hai	45	1	45
Remo—Muni O Muni	55	1	55
Golden Collection—CD Mohd. Raffi Vol. V	275	1	275
Total			375
One Stop Store of Music			With thanks

Figure 6.1(b)

Format of Ledger Entries made at Classic Collection (Figure 6.1(c))

		Ledger			
Date	Qty	Title	Delivery Note/ Receipt No.	Income	Expense
10/8/97	1	HMV Hits—Volume I	1455	45	
	1	Kuch Kuch Hota Hai	1455	55	
	1	Aaja Re Pardesi	1456	35	
	1	Tunak Tunak Tu	1457	225	
	75	Made In India	12567		15,000
	125	Call of Valley—Santoor	12567		6,250
	3	Bad Boys—Micheal Jackson	1458	150	
	1	Made In India	1459	225	
11/8/97	....				
	....				
	....				

Figure 6.1(c)

Stock Record Card for Updating Stock Position (Figure 6.1(d))

Record: Made in India Release Date: 10/10/97		Artist: Daler Mehndi Type: CD, MRP: Rs. 325			
Week Begin 5/5/97 sales	Balance	Week Begin 12/5/97 sales	Balance	Week Begin 19/5/97 sales	Balance
Open stock || |||| ||||| Out of stock +150 |||||||| ||||||||||	11 9 5 0 150 142 132	Open stock |||||| |||||||||| ||||||| |||| ||||||||	132 126 115 108 104 96	Open stock ||||||| ||||||||| |||| |||||| |||||||| +100 |||||	96 89 80 76 70 62 162 157

Figure: 6.1(d)

Special Order Advance Receipt (Figure 6.1(e))

Classic Collection 3, Krishna Towers S.P. Mukherjee Road Fort, Bombay 400 001			
	Special Order: S-1112 Date: 15/10/95		
Customer Name: Lata Chablani Address 301, Raunak Towers Nehru Nagar Kurla East.	Contact No: 5436742		
Title Name East Meet West Jhooth Bole Kauva Kate	Ricky Martin File track	CD Cassette	1 1
Deposit Received: Rs 50	In words:	Rupees Fifty Only	
One Stop Store of Music			With thanks

Figure 6.1(e)

Format of Intimation Letter (Figure 6.1(f))

Dear Ms Lata Chablani
 1. You requested Cassette/CD
 2. East Meet West and Dandia Beets have been received.
Collection before Date: 15/10/95

Yours sincerely,

Bhushan

Figure 6.1(f)

Purchase Order Form for Placing Orders to Suppliers (Figure 6.1(g))

Classic Collection 3, Krishna Towers S.P. Mukherjee Road Fort, Bombay 400 001			
Purchase Order No: HMV/551/97			Date: 15/10/97
To, HMV Records Hero House Fort, Bombay 400 001			
Code	Title and Artist	Type	Nos
WC107	50s Classics—Frank Sinatra	CD	50
KC355	Carnatic Gems by M. Subbalaxmi	Cassette	25
HM545	Doli Saja Ke Rakhna	CD	35
One Stop Store of Music			With thanks

Figure 6.1(g)

Delivery Note as received from one of the Suppliers (Figure 6.1(h))

Seagram Records LTD Unit 17, Riverdale Estate Trafalgar Square, Bombay 400 050			
Delivery Note: D/656			
Against Your Order No. SEA/245/97			
Code	Title and Artist	Type	Nos
WC107	50s Classics—Frank Sinatra	CD	50
KC355	Carnatic Gems by M. Subbalaxmi	Cassette	25
HM545	Doli Saja Ke Rakhna	CD	35

Figure 6.1(h)

Format of Goods Return Note (Figure 6.1(i))

To, Seagram Records LTD Unit 17, Riverdale Estate Trafalgar Square, Bombay 400 050			
Returns Note : R/13/97		Date: 15/10/95	
Code	Title and Artist	Type	Nos
WC107	50s Classics—Frank Sinatra	CD	5
For Classic Collections			

Figure 6.1(i)

6.2 ANALYSIS

The first part of the process of developing an information model for the aforementioned system, i.e. the day-to-day operations at the shop, is to sketch the scope of the processes to be included in the system. This will then translate into a context diagram for the system. We name the system as **Sales and Order system**.

After studying the overview of the current manual system in place for "Classic Collections" and talking to the proprietor Mr. Anirudh Sharma, the following scenario of the system evolved in your mind:

1. Recording details of the sales of CDs and Cassettes made across the counter.
2. Maintaining the details of sales and deliveries so as to know the stock position of cassettes and CDs.
3. To service the special order of cassettes and CDs by customers.
4. Evaluation of stocks and placement of orders to appropriate suppliers.

Combined User-Level Context Diagram for Sales and Order System (Figure 6.2(a))

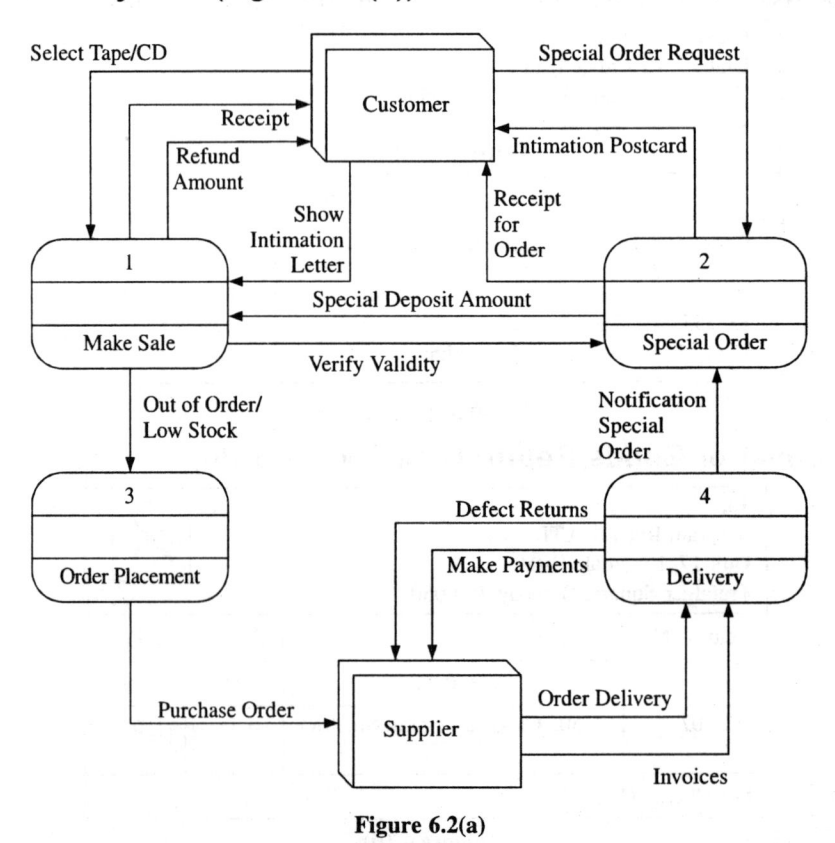

Figure 6.2(a)

From the user-level context diagram in Figure 6.2(a), a preliminary list of basic procedures in the sales order system is made.

1. *Make Sale process* concerns the recording of payments received against sales and refund against cancelled orders. It also verifies the validity of special order by customers at the time of delivery.
2. *Special order process*, processes special orders along with advance, from customers. After recording the special order it reproduces the receipt, and notifies the customer of the arrival on receipt of the cassette/CD. It also updates the details of deposit received to the make sales process when the customer comes to collect and pays the balance amount.
3. *Order Placement* concerns the placing and forwarding of orders to suppliers. It does the analysis of stock in order to determine the quantity of cassettes/CD's to be ordered.
4. *Delivery process* concerns with receipt of deliveries and recording details against order placed. In addition it should be able to identify cassette/CD on special orders and notify the *special order system* of their arrival.

Organization-level Context Diagram (Figure 6.2(b))

Based on the above analysis of each process at the time of system modeling,

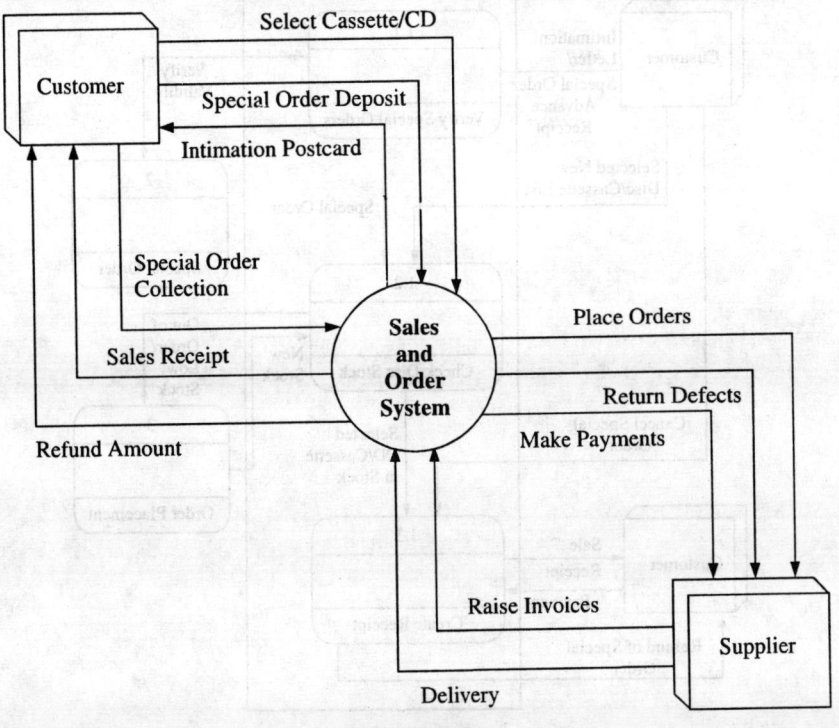

Figure 6.2(b)

the analyst presents the final scope of the system by drawing out the combined context diagram for *Classic Collections*.

Data Flow Diagram

As described in Chapter 2, level-0 DFD describes the system-wide boundaries, detailing inputs to and outputs from the system and major processes. Thus level-0 DFD is similar to the combined user-level context diagram. Hence we take, the user-level context diagram as shown in Figure 6.2(a) as Level-0 data flow diagram.

Thus we have four major processes detailed for the "Sales & Order system". They are:

- make sale
- special order
- order placement
- delivery receipt

Next we will explode each of these processes in Level-1 and Level-2 data flow diagrams.

Level-1 DFD for Make Sale Process (Figure 6.2(c))

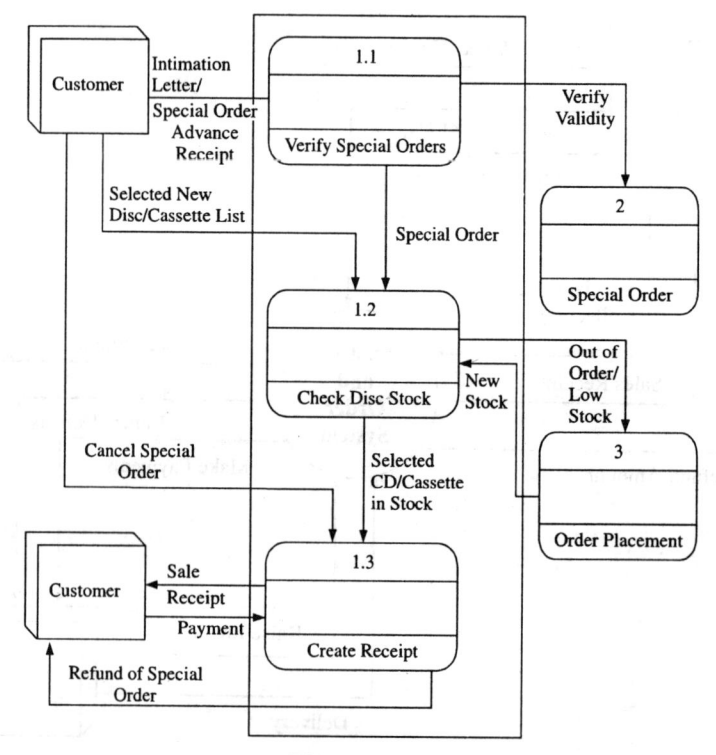

Figure 6.2(c)

Note: The process of selection by customers and entering into list, being a physical process, is out of scope of the system.

In make sales process, we have included the delivery of special orders. When the customer comes with intimation letter or special order receipt, the order is verified to check if the item has been delivered, and also if the customer has come before the stipulated time to collect.

Also, the create receipt sub-process will refund the amount in case the customer does not want the same.

Level-1 DFD for special order process (Figure 2(d))

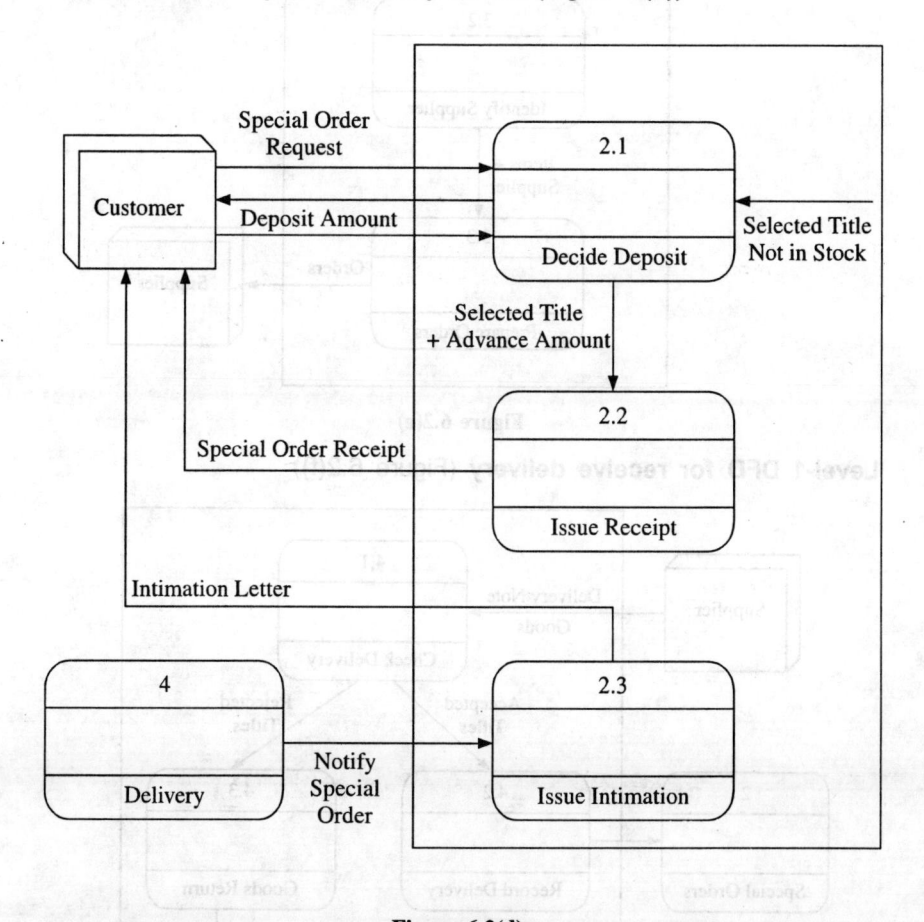

Figure 6.2(d)

From the above DFD, we have included one more activity in the special order process, decide deposit. Before accepting the deposit, stock is checked for availability of the CD.

When the company, with which the order was placed, delivers the items, the store will inform the customer to collect the order.

130 *Workbook on Systems Analysis and Design*

Level-1 DFD for order placement (Figure 6.2(e))

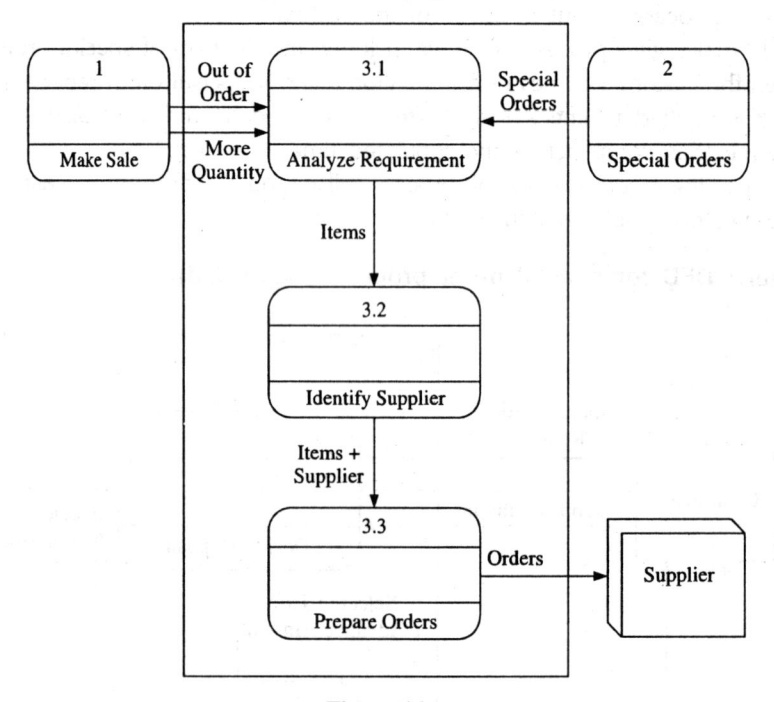

Figure 6.2(e)

Level-1 DFD for receive delivery (Figure 6.2(f))

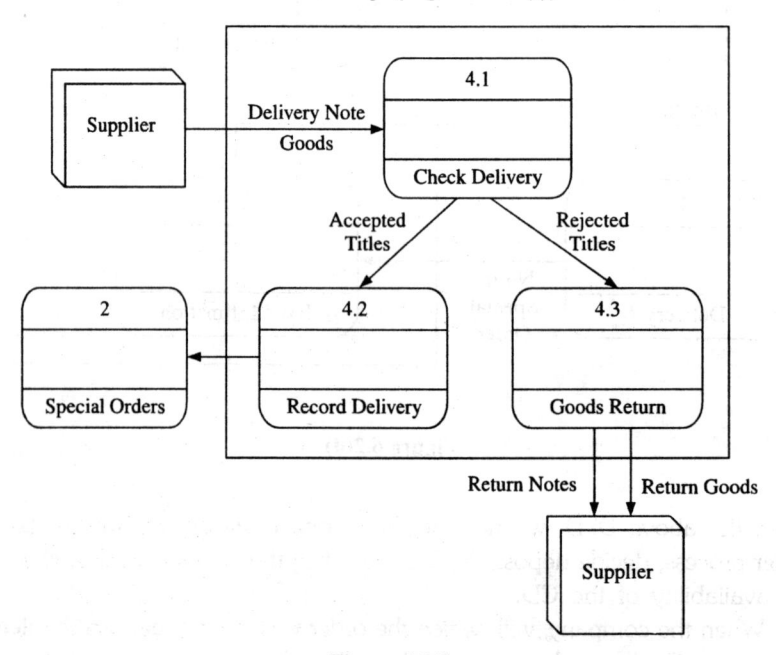

Figure 6.2(f)

Level-2 DFD for make sales—verify special orders (Figure 6.2(g))

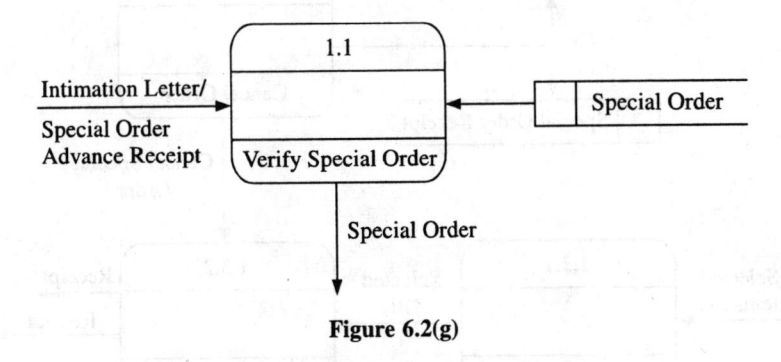

Figure 6.2(g)

When the customer comes with the intimation letter to collect his order, verify special orders, check if the order is not past due date.

Level-2 DFD for make sales—check disc stock

When the customer presents the list of titles he wants to purchase, the titles are checked for their validity and availability in stock (Figure 6.2(h)). If the sales desk finds that a particular item has a low or zero stock level it can immediately place the item in the order list.

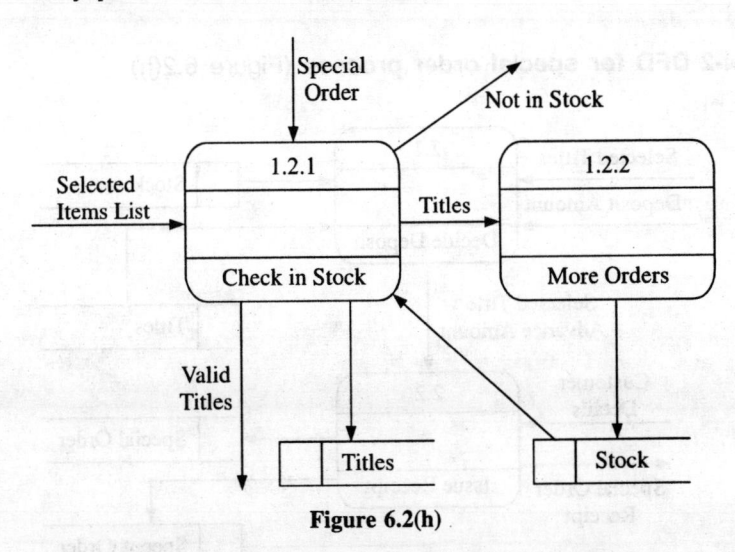

Figure 6.2(h)

Level-2 DFD for create receipt in make sales process

The valid titles are entered and the exact amount to be payable is worked out. When the customer pays, a receipt is produced and the sale is completed (Figure 6.2(i)). In case the customer wishes to get refund of the deposit he made while placing special orders, the same is paid back by updating the special order, and refund is effected after deducting the necessary amount.

132 Workbook on Systems Analysis and Design

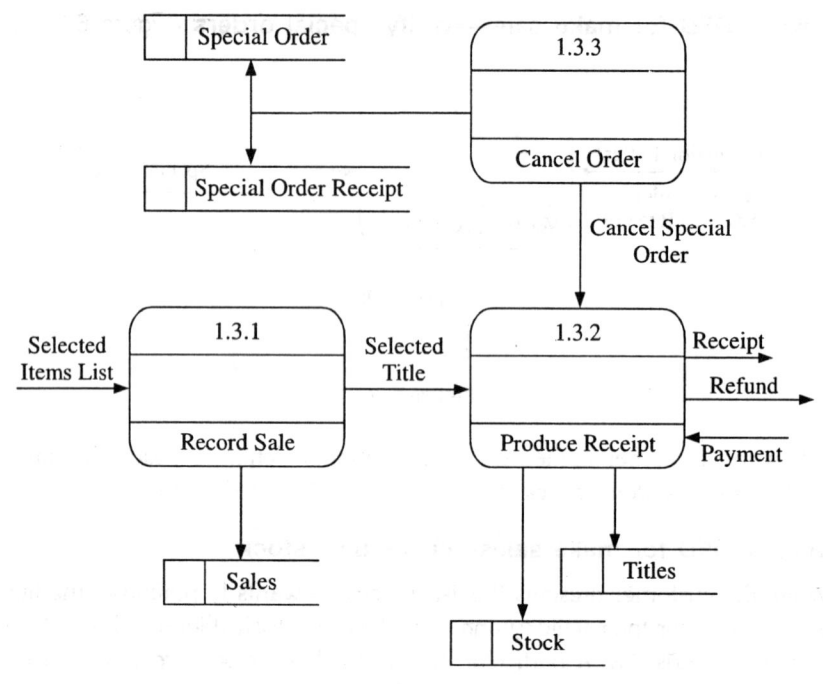

Figure 6.2(i)

Level-2 DFD for special order process (Figure 6.2(j))

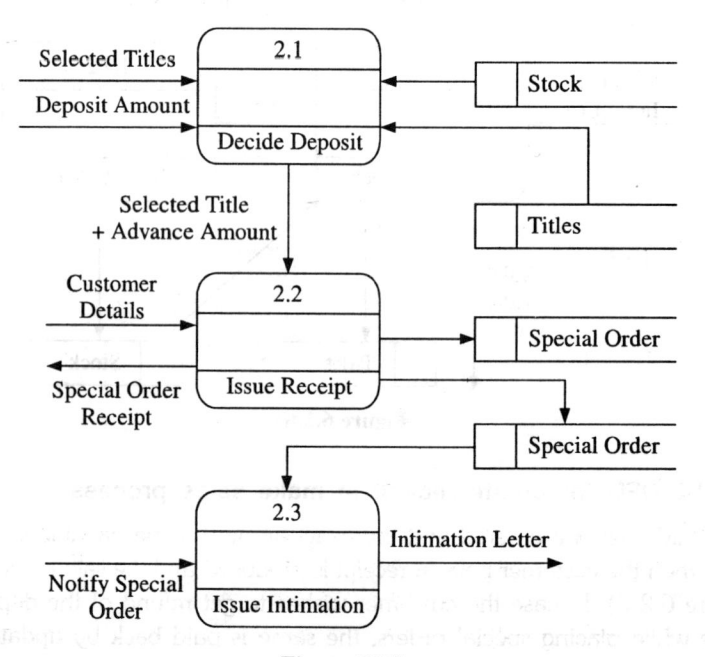

Figure 6.2(j)

Level-2 DFD for order placement process (Figure 6.2(k))

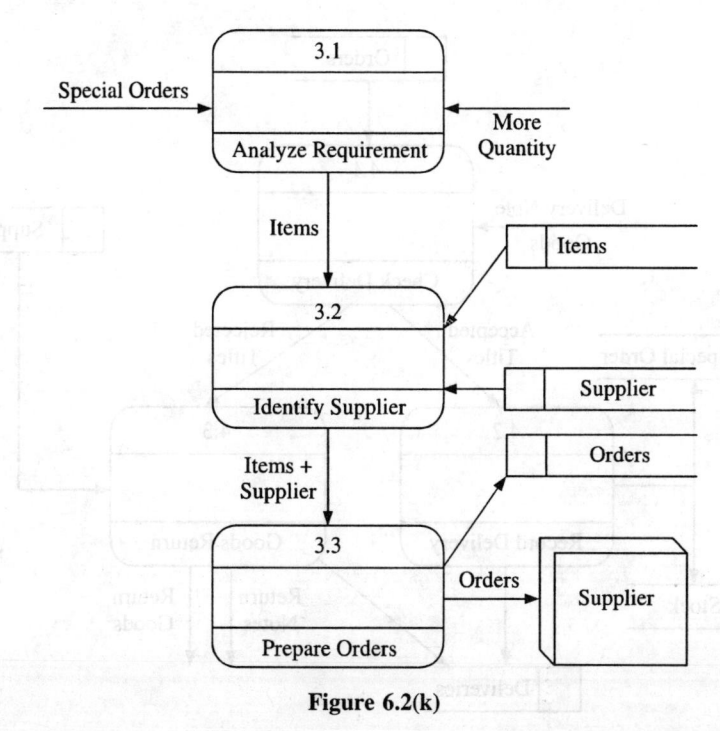

Figure 6.2(k)

Level-2 DFD for Analyze Requirements in Order Placement Process (Figure 6.2(l))

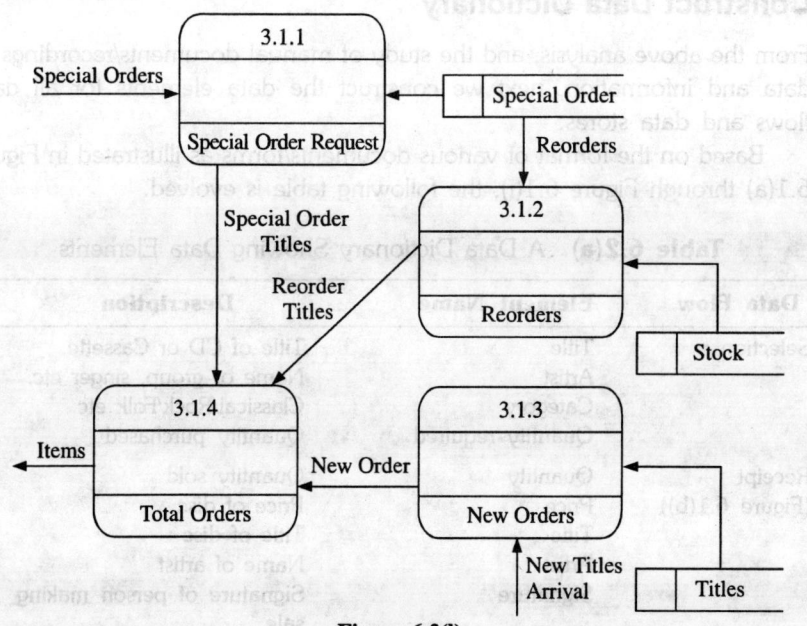

Figure 6.2(l)

Level-2 DFD for Receive Delivery Process (Figure 6.2(m))

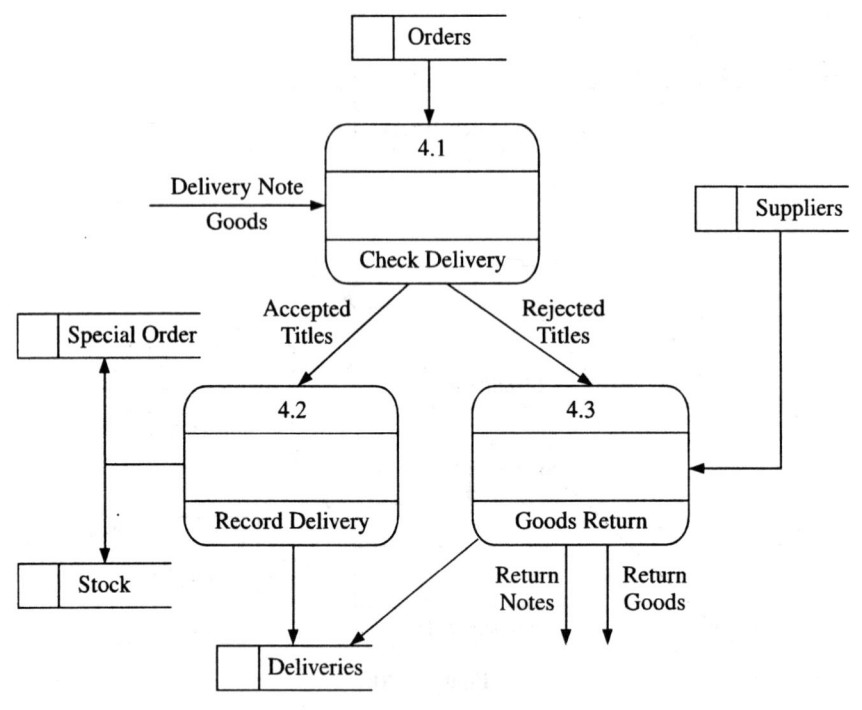

Figure 6.2(m)

Construct Data Dictionary

From the above analysis, and the study of manual documents/recordings of data and information, next we construct the data elements for all data flows and data stores.

Based on the format of various documents/forms as illustrated in Figure 6.1(a) through Figure 6.1(i), the following table is evolved.

Table 6.2(a) A Data Dictionary Showing Data Elements

Data Flow	Element Name	Description
Selection	Title	Title of CD or Cassette
	Artist	Name of group, singer etc.
	Category	Classical/Rock/Folk etc.
	Quantity required	Quantity purchased
Receipt	Quantity	Quantity sold
(Figure 6.1(b))	Price	Price of disc
	Title	Title of disc
	Artist	Name of artist
	Signature	Signature of person making sale

Case Study **135**

Table 6.2(a) (Cont.)

Data Flow	Element Name	Description
Sales	Artist	As above
	Title	As above
	Quantity	As above
Special order receipt (Figure 6.1(e))	Customer name	Customer name
	Customer address	Customer address
	Title	As above
	Artist	As above
	Category	As above
	Quantity	Quantity ordered
	Special order date	Date on which the special order was placed
	Deposit received	Deposit amount received
	Signature	As above
Special order request	Title	As above
	Artist	As above
	Category	As above
	Deposit amount	Amount deposited for Cassette/CD
Intimation Letter (Figure 6.1(f))	Customer name	As above
	Category	As above
	Artist	As above
	Title	As above
	Due date	Date till the order can be collected
Order	Supplier name	Supplier name
	Supplier address	Supplier address
	Artist	As above
	Title	As above
	Disc number	Disc identification no.
	Quantity ordered	Quantity ordered
	PO date placed	Date when order was placed
Delivery (Figure 6.1(h))	Artist	As above
	Title	As above
	Disc number	As above
	Quantity received	Quantity received
Returns (Figure 6.1(i))	Artist	As above
	Title	As above
	Disc number	As above
	Quantity returned	Quantity returned
	Reason	Reason for returning
Special order	Artist	As above
	Title	As above
	Category	As above
	Label Name	As above
Delivery details	Artist	As above
	Title	As above
	Disc number	As above

Table 6.2(a) (Cont.)

Data Flow	Element Name	Description
	Quantity accepted	Quantity accepted = Quantity received − Quantity returned
Stock recording (Figure 6.1(d))	Title	As above
	Artist	As above
	Release date	Date of release of disc
	MRP	Maximum retail price
	Week beginning date	Week beginning date
	Week sales	Number of sales
	Quantity accepted	As above

Based on the documents of *Classic collections*, we explored all the possible data flows and data stores. Next we made a list of provisional data dictionary of the data elements involved in the system as shown in Table 6.2(b).

Provisional Data Dictionary

Table 6.2(b) Provisional Data Dictionary

Element Name	Value Set	Description
Artist	Any characters	Name of group, singer etc
Category	AA	Category of disc e.g. WF = Western Folk, IR = Indian Rock, etc.
Customer add.	Any characters	Customers addresses
Customer name	Any characters	Customers names
Delivery no	Combination of ?	Delivery note number
Deposit recd.	Alphabetic	Y = Received, N = No deposit
Deposit amt.	Numeric	Deposit Amount
Disc no	Combination of ?	Disc identification number
Disc pack	Alphabetic	A = Album, S-Single etc.
Company name	Any characters	Record company
Company add.	Any characters	Address of record company
Order no	Combination of ?	Purchase order number
PO date placed	Date	Date order was placed with supplier
Quantity sold	Numeric	Quantity sold on one sale
Quantity accepted	Numeric	Quantity accepted into stock
Quantity ordered	Numeric	Quantity ordered on one order
Quantity received	Numeric	Quantity of a disc required by a customer
Quantity returned	Numeric	Quantity returned to supplier
Disc type	Character	CA − Audio cassette, CD − Audio CD, VD −Video CD.
Reason	Any characters	Reason for returning discs
Receipt no	Combination of ?	Receipt number
Release date	Date	Date on which disc was released
Return date	Date	Date of return

Table 6.2(b) (Cont.)

Element Name	Value Set	Description
MRP	Amount	Selling price
Signature	Any characters	Person who made sale
Special order date	Date	Date special order was placed
Special offer price	Amount	Special offer price
Supplier name	Any characters	Name of supplier
Supplier add.	Any characters	Supplier address
Title	Any characters	Title of disc
Week begin date	Date	Week beginning date
Weekly sales	Numeric	Total sales for a week

Note: "?" implies the value can be combination of different codes/values.
  All Dates must be stored in YYYYMMDD format.
  All amount value fields must be 8 numeric with 2 decimals.

Next we compile a list of data flows and data stores based on data flow diagrams.

E-R Diagram (Figure 6.2(n))

Define and group attributes for each data entities

Disc : *Disc No* + Title + Artist + MRP + Category + Disc Type + Disc Pack + Company Code + Date of Release

Company : *Company code* + Company Name + Company Address

Stock : *Disc No* + Quantity in Stock + Reorder Level + More Orders (Yes/No flag) + No order

Disc sales : *Receipt No* + *Disc No* + Date + Quantity + Price + Discount + Sale Person ID

Discounts : *Disc No* + Discount Type (Fixed/Percentage) + Discount Per cent + Discount Amount + Discount till Date

Special order : *Sp. Order No* + *Disc No* + Date + Quantity Required + Deposit Amount

Customer : *Sp. Order No* + Customer Name + Address + Contact No

Supplier : *Supplier ID* + Supplier Name + Address + Contact No + Contact Person

Order : *Order No* + Supplier ID + Disc No + Date + Qty Ordered

Order delivery : *DelNote No* + Order No + Disc No + Quantity Delivered

Delivery disc : *DelNote No* + *Disc No* + Quantity Received + Quantity Returned + Return Date + Reason

Note. The data elements, which are underlined are keys to each data store.

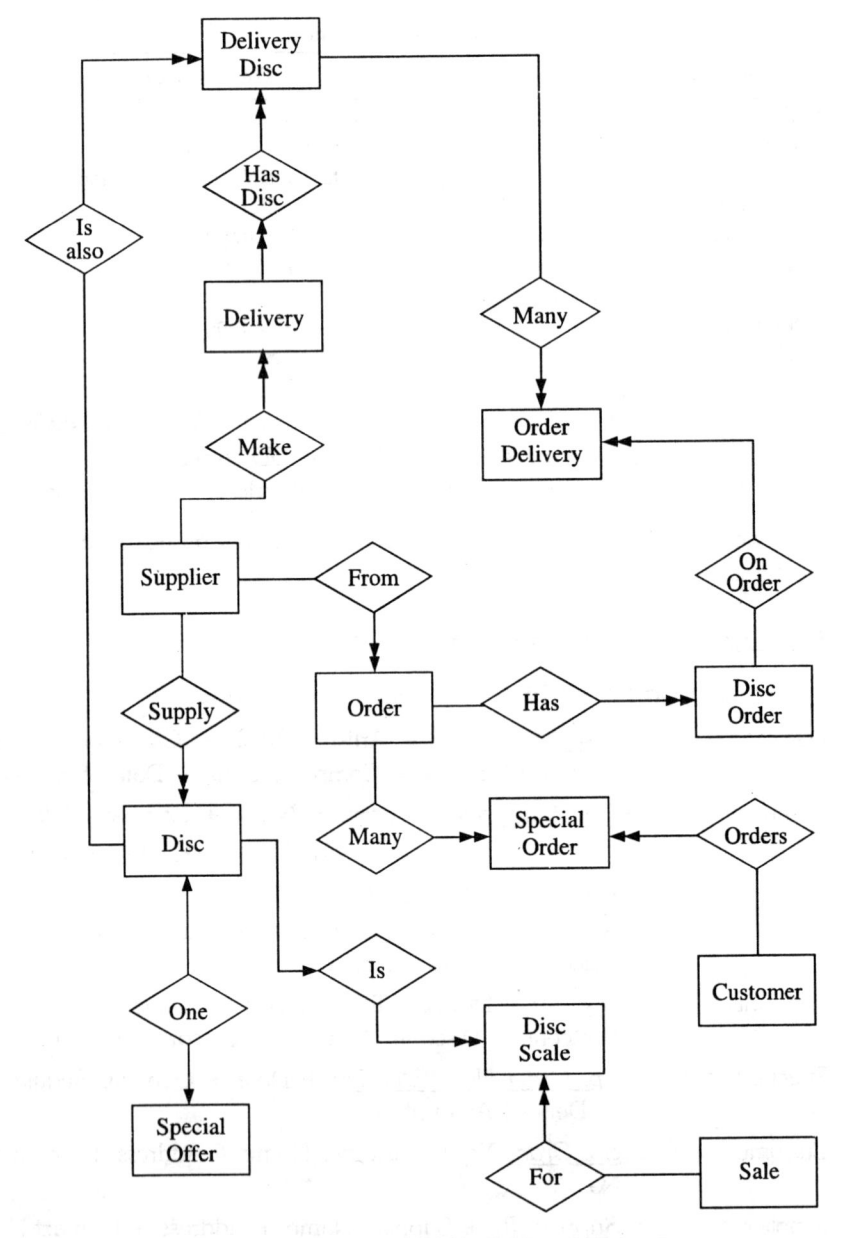

Figure 6.2(n)

6.3 CASE STUDY #2
Overview

Gateway Marina enterprise owns a floating car park for boats on the west-end of the Ganges at Calcutta. Boat owners from far off places bring their wares to the city for trade and park their boats at the floating park. The floating Boat park has *berths (parking spaces)*, these berths are defined by a number of *floating piers* to which they are attached. The floating piers are of varying sizes depending upon the type and size of the boat they can handle. The boats are parked alongside the finger berths, as indicated in Figure 6.3(a). A letter identifies each pier, and each berth has a unique number within the pier.

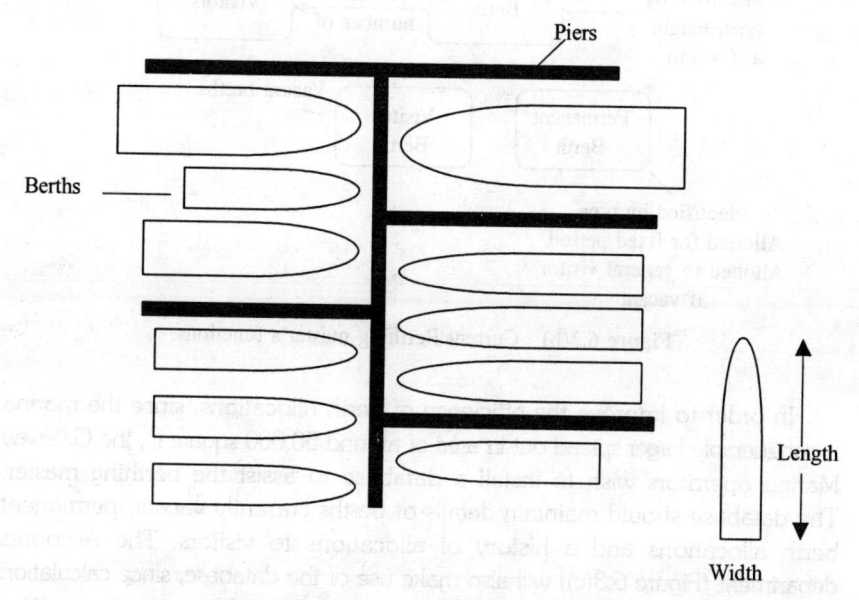

Figure 6.3 (a) View of Gateway Marina.

Each pier has fixed number of berths, each of which is capable of accommodating a boat up to a maximum length and width. These berths are either rented annually to a specific owner for a particular boat or designated as a 'visitors berth'. A given owner may have more than one boat in the marina. Boats have unique names, a length and a beam. Visiting boats are accommodated either in visitor's berths, or in berths which are temporarily vacant because their normal occupants are away. Permanent berth-holders advise the marina office when they will be away so that their berth may be offered to another visitor. Visiting boats may make a number of visits, but usually for not more than one or two nights at a time.

The berthing master's task is to allocate a berth to each visiting boat as it arrives, in such a way that the best use is made of the available berths

140 *Workbook on Systems Analysis and Design*

(Figure 6.3(b)). For this he needs to know which berths are available, and for how long, and the maximum length and beam which they can accommodate. Clearly, a berth cannot be allocated if it is already occupied, or if it will be, i.e. its normal occupant will have returned by that time, before the visitor is due to depart. At the time of booking information is recorded about the owners of visiting boats.

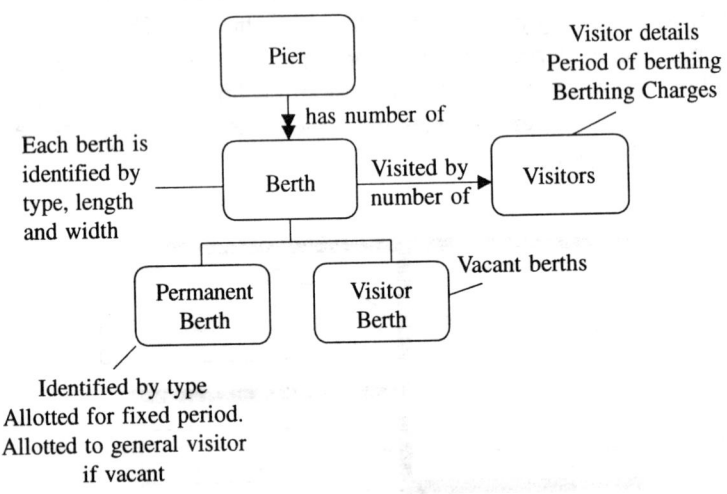

```
                              Visitor details
              Pier            Period of berthing
                              Berthing Charges
           has number of
Each berth is
identified by                 Visited by
type, length      Berth       number of    Visitors
and width

                                       Vacant berths
         Permanent      Visitor
         Berth          Berth

Identified by type
Allotted for fixed period.
Allotted to general visitor
     if vacant
```

Figure 6.3(b) Current Berthing master's functions.

In order to improve the efficiency of berth allocations, since the marina is considerably larger spread out in area of around 50,000 square ft., the Gateway Marina operators wish to install a database to assist the berthing master. The database should maintain details of berths currently vacant, permanent berth allocations and a history of allocations to visitors. The Accounts department (Figure 6.3(c)) will also make use of the database, since calculation of the berth rentals is complex.

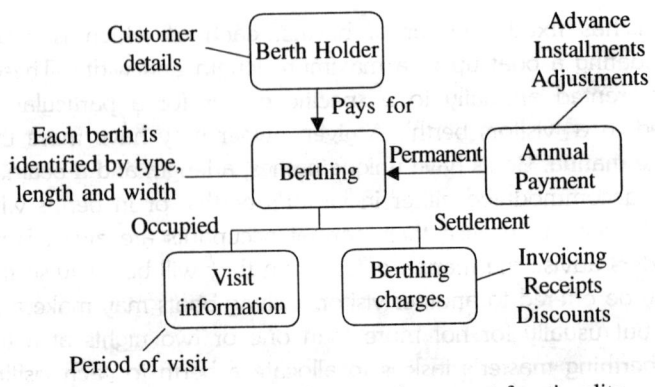

```
                                         Advance
      Customer                           Installments
      details     Berth Holder           Adjustments

                      Pays for
Each berth is
identified by type,             Permanent    Annual
length and width    Berthing                 Payment
       Occupied                 Settlement
                                           Invoicing
              Visit         Berthing       Receipts
              information   charges        Discounts

       Period of visit
```

Figure 6.3(c) Current Accounts department functionality.

Essentially, an annual rental is charged for a berth to permanent berth holders, based on the maximum length and breadth, which it can accommodate. This rental is paid with a single advance payment, or in two equal payments for six months each at a time. In some cases, for certain berth holders the rental is paid in monthly installments. A discount of 10% is offered in the first case, and a handling charge of 10% is added to the total amount if it's a monthly payment to the last installment. At the end of every year, the Marina gives certain discount to the permanent members for their birth being used by visiting boats during the absence of the berth-holder's own boat. This is adjusted against the rental payment/installment for the subsequent year. Apart from this, the marina offers servicing – like workshop repairs, moving the boat from one berth to another, lifting the boat into or out of the water, etc. For servicing additional charge is levied to the visitors.

A permanent member is entitled to one free service during the year. Such costs are invoiced separately and billed to the visitors. The workshop maintains information about boats, which are berthed in Marina and has to be serviced, or which comes to Marina exclusively for repairs. Apart from brief details of each job – date, boat type, description, hours of labour, parts supplied (with cost of each), supervising foreman, total cost – the basic physical details of each boat are also stored–type, length, overall, breadth along with the owner's name and contacts. In addition, the workshop maintains a record of the number, make, size and configuration of the engines are recorded.

Shrikant Shastri who owns Gateway Marina has appointed Cybertech Systems to study, analyze, design and develop a computerized system to handle the above activities for his store. You have been designated as System Analyst by Cybertech Systems to carry out the SSAD for the system. Shrikant wants an integrated system keeping in view the requirements for berthing master, accounts department and workshop. Also in the new system, the Gateway Marina wants to have data analysis done for available berths versus occupancy, revenue generated through workshop, inventories of spare parts and so on.

The company has provided with the current view of the manual system of Gateway Marina, which is illustrated below, in Figures 6.3(c) and Figure 6.3(d).

Based on the above, do the following in this order applying SSAD techniques.

(a) **Evolve a context diagram that summarizes the information flow for Gateway between berthing, accounts and workshop departments.**

(b) **Convert the context diagram into DFD to explain the process and data flows for each sub-system.**

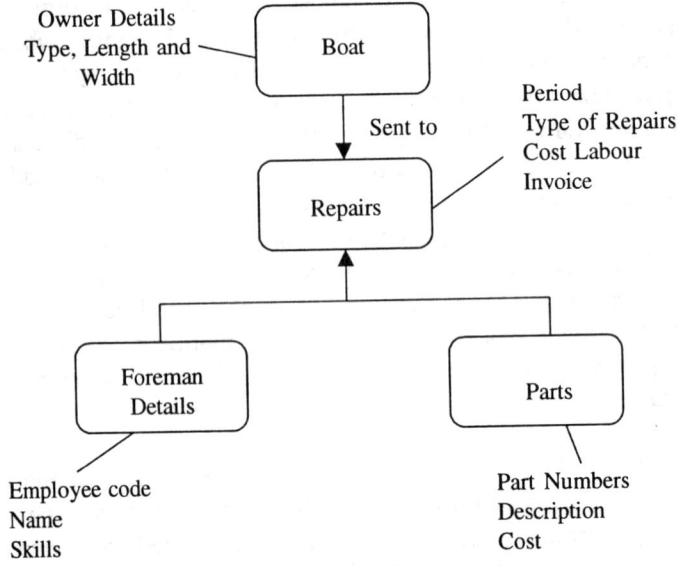

```
                Owner Details
                Type, Length and ──┐  ┌─────────┐
                    Width          │  │  Boat   │
                                   └──┤         │
                                      └────┬────┘      Period
                                           │ Sent to   Type of Repairs
                                           ▼           Cost Labour
                                      ┌─────────┐      Invoice
                                      │ Repairs ├─────
                                      └────┬────┘
                                           ▲
                         ┌─────────────────┴─────────────────┐
                   ┌─────────┐                         ┌─────────┐
                   │ Foreman │                         │  Parts  │
                   │ Details │                         │         │
                   └─────────┘                         └─────────┘
```

Employee code   Part Numbers
Name   Description
Skills   Cost

Figure 6.3(d)   Workshop Functionality.

(c) **Based on the DFD output, list out the entity relationships between data stores and normalize the data structures.**

(d) **Convert the processes in DFD into structure charts.**

(e) **Use the process structure charts to work out the process specifications.**

7
Object-Oriented Analysis and Design

7.1 SSAD vs OOAD

The three important design methods, which have evolved and had played dominating roles in systems analysis, design and development, are:

- Top-down structured design, also known as SSAD (Structured System Analysis and Design)
- Data-driven design
- Object-oriented design

Each of these variations applies algorithmic decomposition, but the most influential of the three has been top-down structured design. SSAD method was influenced by the topology of traditional high-order programming languages, like COBOL, FORTRAN, etc. In these languages, the fundamental unit of decomposition is the sub-program, and the resulting program is where these sub-programs perform their work by calling other sub-programs. **To sum up, SSAD applies algorithmic decomposition to break a large problem down into smaller steps.**

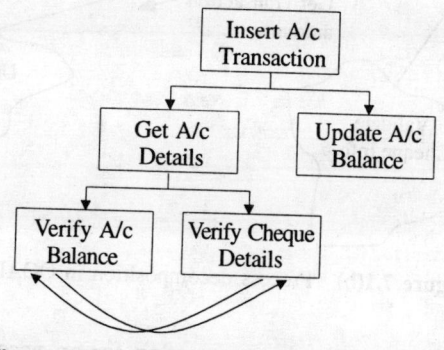

Figure 7.1(a) Process decomposition in SSAD.

SSAD does not measure well for extremely complex systems, where the structure of a software system is derived by mapping system inputs to outputs. Data-driven design has been successfully applied to such systems, particularly management information systems, which involve direct relationships between the inputs and outputs of the system, and is less concerned for time-critical systems. **In Data-driven analysis and design method of the structure of the system being developed are derived by mapping system inputs to outputs.**

But with the advent of high-order programming languages such as Small talk, C++ etc., it was possible to lay emphasis on modeling of software systems as collection of cooperating objects, treating individual objects as instances of a class within a hierarchy of classes. **Thus in OOAD, the emphasis on building applications is not top-down, but bottom-up, where by software becomes a collection of discrete objects that incorporate data structure and behaviour.**

Designing OO Systems

When designing an object-oriented (OO) based software system, it is essential to decompose the systems into smaller and smaller parts, each of which we may then refine independently.

Example

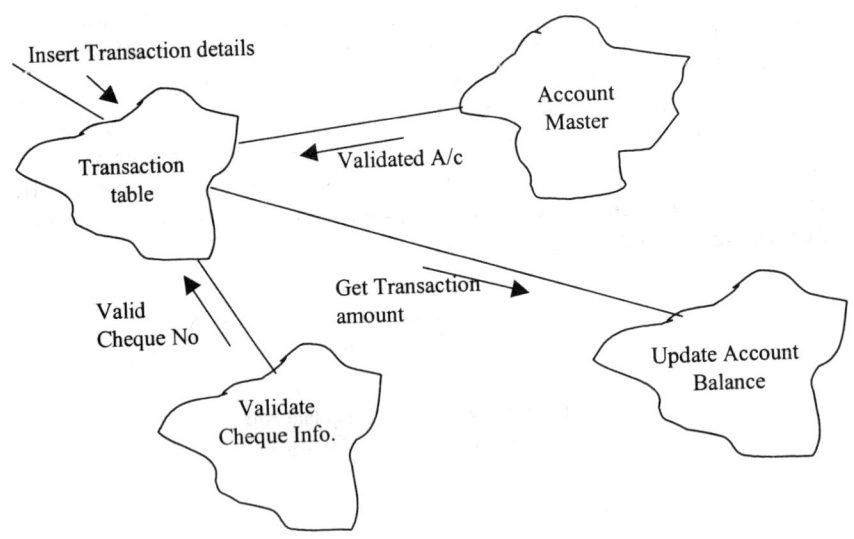

Figure 7.1(b) Process decomposition in OOAD.

Figure 7.1(a) illustrates how a transaction for an account is updated in a bank. It follows the procedural nature of carrying out the task by way of structure charts of SSAD. The same, when designed using object-oriented, takes the form shown in Figure 7.1(b). In this, the **transaction table, account master, update account balance, validate cheque information are objects on which messages are sent to carry out the required operations**. Each of these is built up of individual objects from which the information is inherited.

Example

Object *validate cheque information*, may call objects *valid account* and *valid cheque number*, to get the cheque information. Similarly when the object *transaction table* receives the account transaction details to be inserted, while inserting, it validates for account and cheque number by using objects *valid account* and *valid cheque number*. Further, the object *update account balance*, adds the transaction amount to account balance or subtracts the transaction amount from account balance based on the message service. Next time, suppose if there are changes to the amount of information that is being inserted into the transaction table, what changes is the message; insert transaction details and the associated object; transaction table, instead of rewriting the full program logic.

This way one can understand and comprehend the functionality to bring changes/enhancements, and to this we need to only look at small portion of design rather than all parts at once. **From the above it is very clear that, SSAD highlights the ordering of events, and the OOAD view emphasises the agents that either initiate/cease action or are the subjects upon which these operations act.**

Contrasting Object and Traditional Systems

Figure 7.1(c) represents how an object system is made up of inter-linking objects. Compare this to a traditional system of application development as shown in Figure 7.1(d). In traditional systems, one constructs individual programs to process a set of information, which is retrieved and updated to a common database. Thus any change in the database structure may lead to changes in one or more programs, hence leading to change blues.

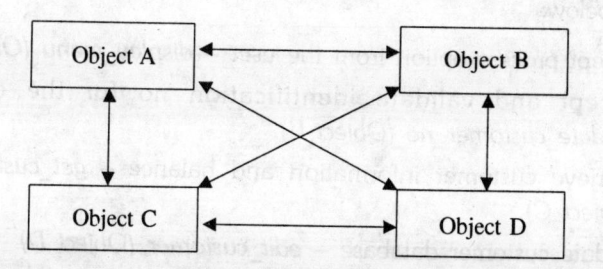

Figure 7.1(c) An object system is a set of collaborating objects.

146   Workbook on Systems Analysis and Design

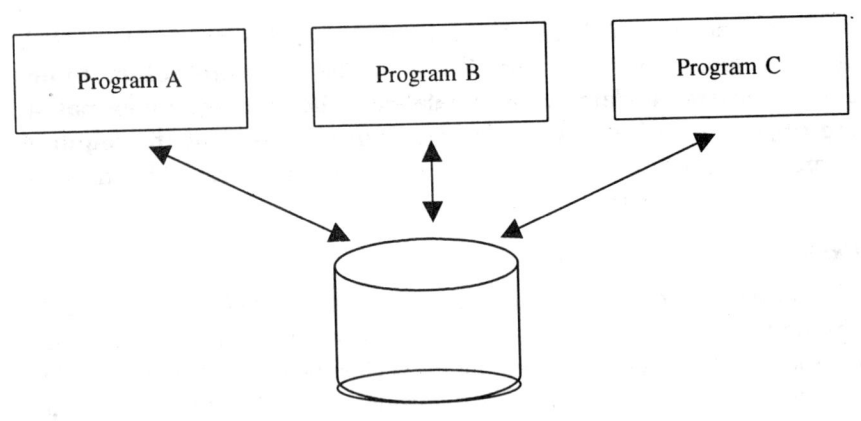

Figure 7.1(d)   An traditional system with programs and data files.

On the other hand, in the case of object programming each class acts on its own set of data, or in other words, to each set of data are associated its object classes, which encapsulates the way data needs to be handled. It's through the use of these objects that the data are shared, instead of the database being openly accessible by one and all as in the case of traditional systems.

In Figure 7.1(d), Program A is to edit customer information in an application, and consists of many paragraphs as described below:

- The *first paragraph* is coded to accept process option from the user.
- The *second paragraph* is to accept and validate customer identification no.
- In the *third paragraph* you write logic for retrieving the customer information and customer balance from the customer database.
- Finally in the *fourth paragraph* you commit input data into database.

From Figure 7.1(c), in OO, the correspondence to program A is written in the following way. Each of the four paragraphs corresponds to module as described below:

- Accept process option from the user – display_menu (Object A).
- Accept and validate identification no for the customer – validate_customer_no (Object B).
- Retrieve customer information and balance – get_customer_info (Object C).
- Update customer database – edit_customer (Object D).

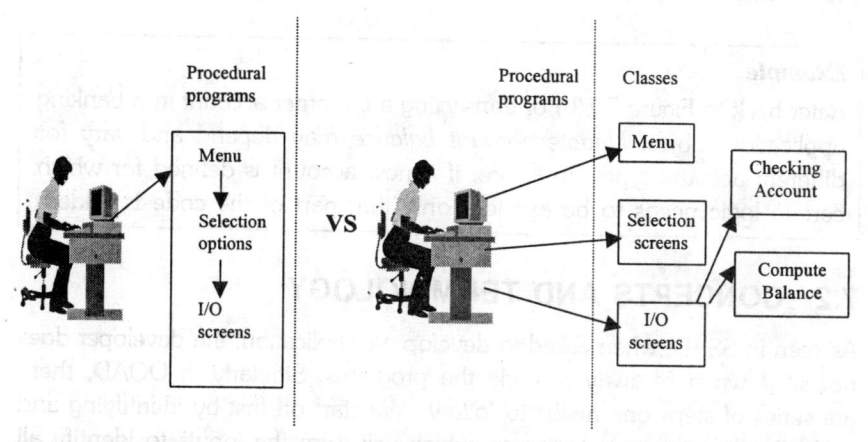

Figure 7.1(e) Programmatic view of traditional application versus object applications.

To sum up, in OO, the physical building block is the module, which represents a logical collection of classes and objects instead of subprograms.

What is Object-Orientation

Object-orientation is not a data modeling approach like E-R diagram, instead object-orientation is an application development philosophy that transcends specific languages, products, and database management systems.

From the standpoint of procedural languages, object-oriented languages also require thinking non-procedurally in much greater degrees of abstraction and actually relearning how to program – i.e. how to attack and solve a set of logical problems.

Advantages of OOAD

- Object-orientation approaches the process of analysis, design and implementation with a problem in terms of application development.

Example

Refer to Figure 7.1(b) of transacting a customer account in a banking application, each of the updation processes of transaction table, – *Account master validation, Update account balance* and *Update Cheque Information* – each is treated separately in terms of its implementation.

- Object-orientation promotes reuse of objects when new business logic is applied to the applications. This is very important for two reasons, firstly modifications required are carried out faster, and secondly it reduces errors and maintenance problems.

> **Example**
> Refer back to Figure 7.1(b) of transacting a customer account in a banking application, object *Update account balance* may depend and vary for different account types. In future, if a new account is defined for which certain logic needs to be applied, only that part of the code is added.

7.2 CONCEPTS AND TERMINOLOGY

As seen in SSAD, when asked to develop an application, the developer does not sit down right away to code the programs. Similarly in OOAD, there are series of steps one needs to follow. We start off first by identifying and understanding the new concepts, which will form the inputs to identify all the components that go into developing an object-oriented applications.

SSAD lays emphasis on tools and techniques to be applied to system design by way of top-down approach, whereas OOAD with **objects and classes** analysis proceeds from bottom up. Thus OOAD is very closely tied to programming, as compared to SSAD which stresses on system analysis and design.

In SSAD, a program consists of modules to meet functional requirements. Functions by way of procedures are the primary focus of this approach and they process/share data with files. In OOAD, program consists of interrelated classes of objects which have data and procedures encapsulated within it.

Thus in SSAD, emphasis is on process modeling while in OOAD, emphasis, in addition, is also on data modeling.

> **Example**
> We don't think of our cars/vehicles as a list of tens and hundreds of incomprehensible parts, instead think of it as a well-defined object with its own unique behaviour to perform in an integrated manner. This notion of abstraction allows one to drive a car/vehicle without being bothered about its complexities of the algorithm that builds it. When looked from outside, the car is a car, but once inside, one can operate the steering, ac, music, cellular phone all at once.

Objects

Objects are a programmatic representation of a real-world entity – for example, an employee, a purchase order, an invoice, a payment. **In traditional terms objects are data structures or record definitions. Object instances refer to physical data defined by these data structures – the actual records or rows.**

> **Example**
> Object *Employee* can have instances ***R K Gupta 12 Cybertech House, Wagle Estate, Thane***
> The object purchase order will have instance ***PO # 235678***, the object *Payment* will have instance as ***$ 555***, and so on.

From Figure 7.2(a), customer object is customer data surrounded by operations (which in OO terms are referred to as methods). The data represents the *attributes* of the object, and the code that surrounds the data, represents the *methods* of the object. In customer object, the *attributes* of the customer object are *name, address, account-no, balance-due*. The methods associated with the customer object are Get-Balance-Due, Adjust-Balance, Send-Sales-Material, Send-Invoice.

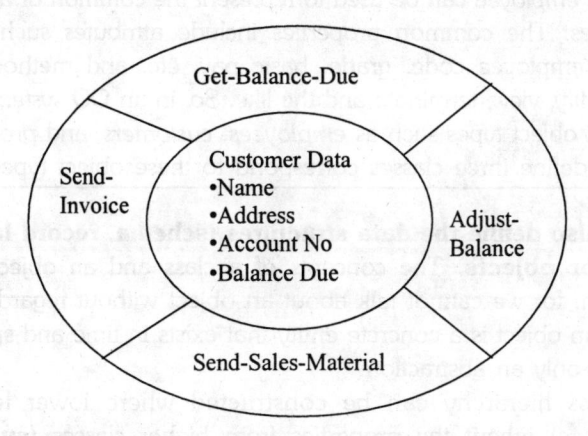

Figure 7.2(a) Conceptual view of a customer object.

We see that objects exhibit behaviour in which each object does certain activity, go through certain stages and so on.

> **Example**
> - Loans in a banking application may be processed, mortgaged, overdue, penalized or paid off.
> - Window objects in a GUI application may get focus or lose focus.
> - Patient in a hospital management application may be registered, checked, admitted, treated, discharged and billed for services.

Inanimate objects, including many business-oriented objects such as *purchase orders, balances* and *payments* in sales application have less obvious behaviours. Business objects are often easy to understand but harder to visualize in object terms. A *purchase order* may be complete, incomplete or overdue, similarly *balances* are positive or negative, *payments* are on time or late and so on.

Classes

In the OO world what takes the place of program development or editing a piece of application code with a definite functionality is the assembly of new applications from pre-existing components called **classes**. In traditional programming terms, an OO class is a complete mini-application consisting of the syntax or methods, the data structure definitions or instance variables, and the physical data or object instances on which methods act.

> *Example*
> The class *employee* can be used to represent the common of all company employees. The common properties include attributes such as name, address, employee code, grade, basic pay, etc. and methods such as add, modify, view, terminate and the like. So, in an OO system you have identified object types such as *employees*, *customers*, and *products*, then you will define three classes correspond to these object types.

Classes also define the data structures (schema, record layouts, and so on) for objects. The concepts of a class and an object are tightly interwoven, for we cannot talk about an object without regard to its class. Whereas an object is a concrete entity that exists in time and space, a class represents only an abstraction.

A class hierarchy can be constructed where lower level classes **(subclasses)** inherit the properties from higher classes **(super class)**. Figure 7.2(b) illustrates a class hierarchy. We can set up a class hierarchy for our *employee* class with subclasses such as 'contract employee' and 'permanent employee'. Similarly class *permanent employee* has subclasses such as 'officers', 'support staff,' etc.

> *Example*
> As illustrated in Figure 7.2(b) the class *officer* is a subclass of *permanent employee*, which is a subclass of *employee*, which in turn is a subclass of *person*.

Finally, a Class can be defined as a set of objects that share a common structure and a common behaviour. Objects that share no common structure and behaviour cannot be grouped in a class. When objects are similar to one another in functions and other attributes, they can be put together into a class.

Object-Oriented Analysis and Design **151**

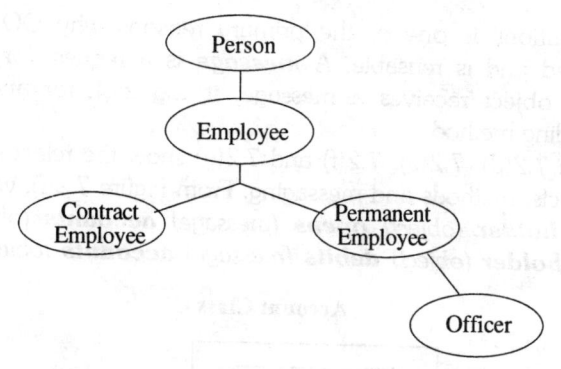

Figure 7.2(b) Object-oriented class hierarchy.

Messaging

In traditional system the data are contained in files, and we develop programs to access these files. However, an object owns and controls its data. The only way one object can access another object's data is to send that object a *message* requesting the data which are effectively hidden or *encapsulated.*

Figure 7.2(c) illustrates two messages sent to the customer object. Each method of an object knows the list of messages to which it can respond and

"Customer (Ra$_j$): Get-Balance-Due (Date= 21022000)"

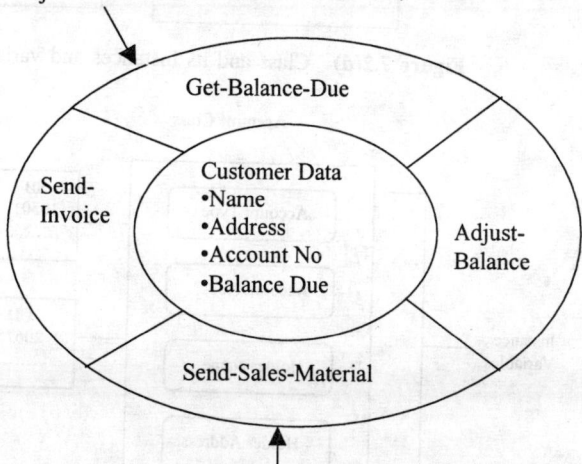

"Customer (Raj), item details (xyz,50,kg,565): Send-Sales-Material (Challan No=145)"

Figure 7.2(c) Object-oriented class hierarchy.

also how it will respond to each of the messages. It's very important to know that the object that receives the messages is responsible for providing the code that is needed for the invoked object.

Thus we see that objects communicate with one another through messages. This messaging concept is very similar to the traditional notion of CALL in procedural languages. This formal process of message

152 Workbook on Systems Analysis and Design

communications is one of the primary reasons why OO code can be modularized and is reusable. A **message** is a request for an operation. When an object receives a message, it will look for and execute the corresponding method.

Figure 7.2(d), 7.2(e), 7.2(f) and 7.2(g) show the relationships between class, objects, methods and messaging. From Figure 7.2(f), we can arrive as **account holder** *(object)* **opens** *(message)* **accounts** *(object)*. Similarly **account holder** *(object)* **debits** *(message)* **accounts** *(object)*, and so on.

Account Class

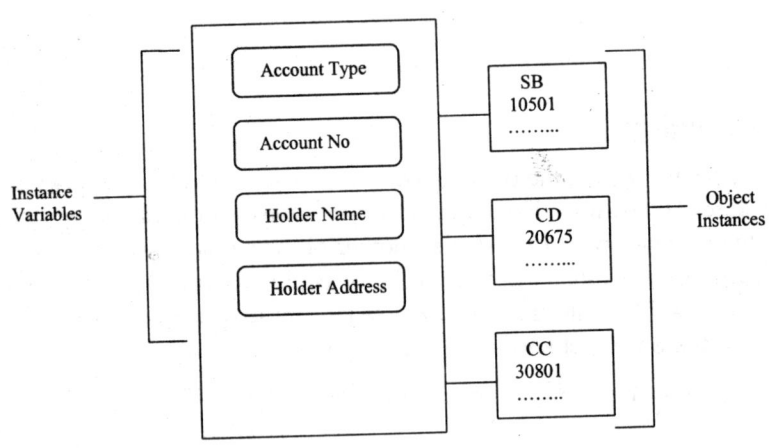

```
Instance Variables:
  Account Type
  Account No
  Holder Name
  Holder Address

Object Instances:
  SB 10501 ........
  CD 20675 ........
  CC 30801 ........
```

Figure 7.2(d) Class and its instances and variables.

Account Class

```
Instance Variables:
  Account Type
  Account No
  Holder Name
  Holder Address

Methods:
  Opening
  Closing
  Transfer

Object Instances:
  SB 10501 ........
  CD 20675 ........
  CC 30801 .....
```

Figure 7.2(e) Methods, which act on objects.

Object-Oriented Analysis and Design **153**

Account Class

Instance Variables — Account Type, Account No, Holder Name, Holder Address

Object Instances — SB 10501 ...., CD 20675 ...., CC 30801 ....

Methods — Opening, Closing, Transact

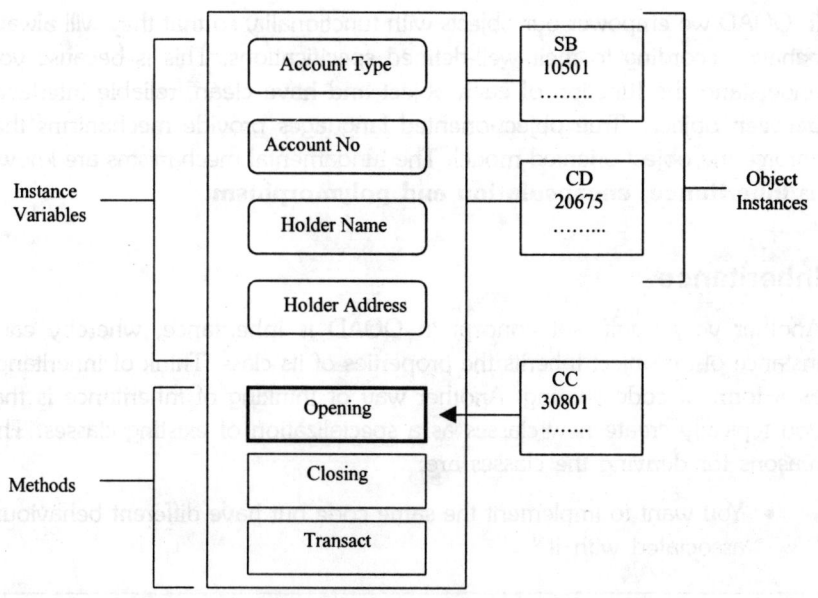

Message (Open account "CC 30801")

Figure 7.2(f) Messaging on object classes.

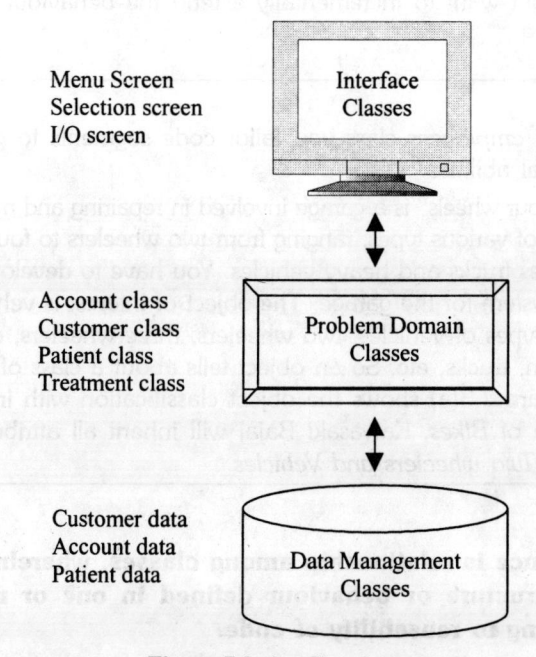

Interface Classes — Menu Screen, Selection screen, I/O screen

Problem Domain Classes — Account class, Customer class, Patient class, Treatment class

Data Management Classes — Customer data, Account data, Patient data

Figure 7.2(g) Class layers.

7.3 OBJECT-ORIENTED THEMES

In OOAD we empower our objects with functionality so that they will always behave according to their well-defined specifications. This is because you understand the function of each object and have clean, reliable interfaces between objects. True object-oriented languages provide mechanisms that enforce the object-oriented model. The fundamental mechanisms are known as **inheritance, encapsulation and polymorphism.**

Inheritance

Another very significant concept in OOAD is inheritance, whereby each instance of an object inherits the properties of its class. Think of inheritance as a form of code sharing. Another way of thinking of inheritance is that you typically create new classes as a specialization of existing classes. The reasons for deriving the classes are:

- You want to implement the same code but have different behaviours associated with it.

> *Example*
> A nurse class might be treated differently than a doctor class, even though they all belong to the class *hospital employees*.

- You might want to incrementally extend the behaviour of the base class.

> *Example*
> - To *hospital employees* class you tailor code segments to give *doctor* class special abilities.
> - "Balance your wheels" is a garage involved in repairing and maintenance of vehicles of various types, ranging from two wheelers to four wheelers, that includes trucks and heavy vehicles. You have to develop a vehicle servicing system for the garage. The object of interest is vehicle. There are many types of vehicles, two wheelers, three wheelers, cars, jeeps, pick up van, trucks, etc. So an object tells about a class of real-world things. Figure 7.3(a) shows the object classification with inheritance. An instance of *Bikes*, Kawasaki Bajaj will inherit all attributes above it, namely *Two wheelers and Vehicles*.

Thus inheritance is relationship among classes, wherein one class shares the structure or behaviour defined in one or more other classes, leading to reusability of code.

Example

From the Figure 7.3(a), an instance of a *150 CC 2 wheeler*, a Hero Honda motorcycle *inherits* all attributes above it in the hierarchy, i.e. *Two Wheelers* and above it *Vehicle*. Thus Hero Honda is a land vehicle, which runs on fuel and is electronically driven with manual leg gears and has 150 cc HP engine.

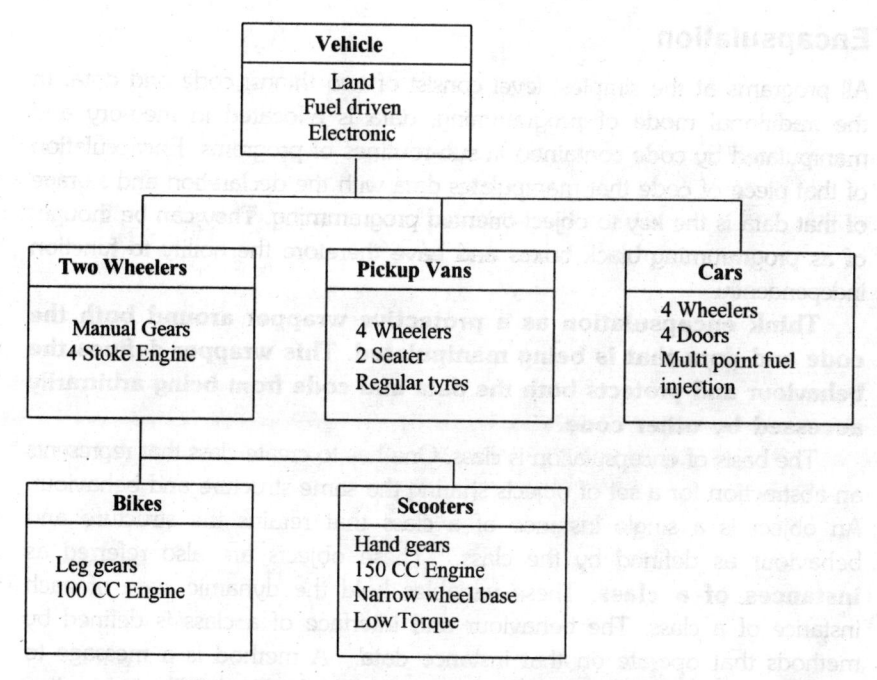

```
                         ┌──────────────┐
                         │   Vehicle    │
                         │   Land       │
                         │   Fuel driven│
                         │   Electronic │
                         └──────┬───────┘
          ┌─────────────────────┼─────────────────────┐
┌─────────────────┐   ┌─────────────────┐   ┌──────────────────┐
│  Two Wheelers   │   │  Pickup Vans    │   │      Cars        │
│                 │   │                 │   │  4 Wheelers      │
│  Manual Gears   │   │  4 Wheelers     │   │  4 Doors         │
│  4 Stoke Engine │   │  2 Seater       │   │  Multi point fuel│
│                 │   │  Regular tyres  │   │  injection       │
└────────┬────────┘   └─────────────────┘   └──────────────────┘
    ┌────┴─────┐
┌─────────────┐ ┌──────────────────┐
│   Bikes     │ │    Scooters      │
│             │ │  Hand gears      │
│  Leg gears  │ │  150 CC Engine   │
│  100 CC Eng │ │  Narrow wheel base│
│             │ │  Low Torque      │
└─────────────┘ └──────────────────┘
```

Figure 7.3(a) Object classification with inheritance.

Data abstraction

OO is similar to traditional development methods because it employs some fundamental principles of good software engineering such as decomposing a problem into smaller, manageable modules and restricting data access. OO encourages modularization and requires restricted data access. However, we still must write the code to define data, and we must code to process that data.

Finally, objects exhibit behaviour in that each object does or is supposed to do certain things, go through certain stages or processes. In Figure 7.2(f) we see that accounts will be opened, closed or transacted. These were termed as **methods**, and they are the ones which define object behaviours. This process of identifying the necessary information about a component is referred to as **Data abstraction**. Given a task, to accomplish we need the characteristics, and in addition, the actions that are to be performed. The advantages of data abstraction are:

- It focuses on the problem
- It identifies the essential characteristics and the actions required
- It helps eliminate unnecessary detail, otherwise one must have got tagged to a particular object.

Encapsulation

All programs at the simplest level consist of two things: code and data. In the traditional mode of programming, data is allocated in memory and manipulated by code contained in sub-routines of programs. Encapsulation of that piece of code that manipulates data with the declaration and storage of that data is the key to object-oriented programming. They can be thought of as programming black boxes and have therefore the ability to function independently.

Think encapsulation as a protective wrapper around both the code and data that is being manipulated. This wrapper defines the behaviour and protects both the data and code from being arbitrarily accessed by other code.

The basis of encapsulation is class. One has to create class that represents an abstraction for a set of objects sharing the same structure and behaviour. An object is a single instance of a class that retains the structure and behaviour as defined by the class. These objects are also referred as **instances of a class**. These variables hold the dynamic state of each instance of a class. The behaviour and interface of a class is defined by methods that operate on that instance data. A method is a message to take some action on an object. These messages look a lot like subroutine calls in older procedural languages.

Thus we see that an important characteristic of OOAD is communication through messages among different objects. These messages may be in the form of services that are performed for the objects.

Figure 7.3(b) shows an example of objects communicating through a message. At time t the object Maruti Zen is received in garage and the parts are broken up by the mechanic. After time $t + 1$ the mechanic object sends the message to the Maruti Zen object indicating that the car is repaired. **As you could see this message activates the method or procedure to change the attribute of Maruti Zen object. Since the routine that makes the change is stored with the Maruti Zen object, it is possible to change other parts of the system without changing this object.**

Further if you change some parts of the system, but continue to send messages in the same format as you did before the changes, there is no need to modify the Maruti Zen and other automobile objects in its class.

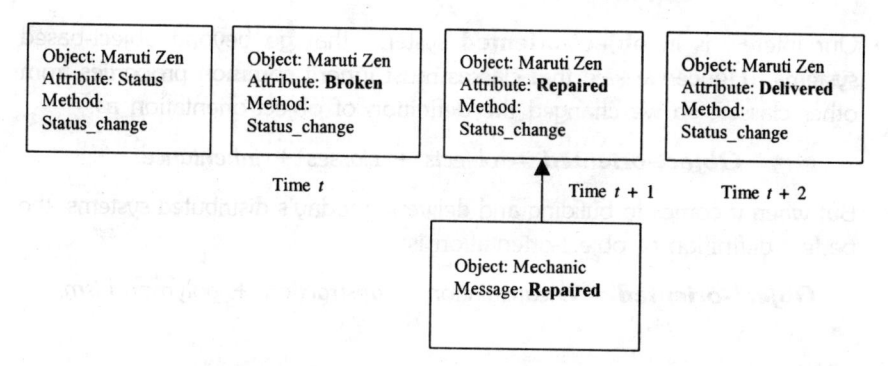

```
┌─────────────────┐  ┌─────────────────┐  ┌─────────────────┐  ┌─────────────────┐
│ Object: Maruti Zen│  │ Object: Maruti Zen│  │ Object: Maruti Zen│  │ Object: Maruti Zen│
│ Attribute: Status │  │ Attribute: Broken │  │ Attribute: Repaired│  │ Attribute: Delivered│
│ Method:         │  │ Method:         │  │ Method:         │  │ Method:         │
│ Status_change   │  │ Status_change   │  │ Status_change   │  │ Status_change   │
└─────────────────┘  └─────────────────┘  └─────────────────┘  └─────────────────┘
         Time t                              ▲    Time t + 1        Time t + 2
                                             │
                                  ┌─────────────────┐
                                  │ Object: Mechanic│
                                  │ Message: Repaired│
                                  └─────────────────┘
```

Figure 7.3(b) Message passing and encapsulation in object-oriented systems.

Polymorphism

In most of the programming languages, in order to complete two different tasks, one has to have two separate functions with different names. Polymorphism, which means one object taking many shapes, is a simple concept that allows a method to have multiple implementations.

Polymorphism is the ability of objects to react differently to an identical message. An Object reacts differently based on the information supplied, and understands the context of the information that has been input.

Example

In procedural programming languages, variables and values can be interpreted to be of one and only one type, but with object-oriented languages some variables and values may be more than one type.
Date conversion function is a typical example of polymorphism.
Inputs: Date, Format (YMD or DMY)
Output: Converted Date
In this function you send a date along with the format which can be either YMD or DMY. If the input date is in YMD format, then the function converts the date to DMY format and vice versa.

This way you see that the same function responding and delivering the desired result to different parameters.

Object-Based vs Object-Orientation

The basic object concepts can be used to represent powerful programming systems. An object-based system is defined as follows:

$$\textbf{Object-based} = objects + classes$$

Our interest is in **object-oriented** systems that go beyond object-based systems. Hence we said that classes must inherit common properties from other classes, so we changed the definition of object-orientation as:

Object-oriented = *objects + classes + inheritance.*

But when it comes to building and delivering today's distributed systems, the perfect definition of object-orientation is:

Object-oriented = *encapsulation + abstraction + polymorphism.*

Conclusion

Object-oriented analysis and design is the method that leads us to an object-oriented decomposition. By applying OO design, we create software systems that is resilient to change and written with economy of expression. It also offers a rich set of logical and physical models with which we may reason out different aspects of system under consideration. These models are classified as Object-Oriented Analysis (OOA), Object-oriented Design (OOD) and Object-oriented Programming (OOP).

One of the benefits of using objects is data encapsulation. The data in an object is accessible only through that object's methods. This protects the data from corruption. Further, Inheritance provides several benefits. First, it enables the reuse of attribute and method definition and secondly it facilitates maintenance.

7.4 HOW TO IDENTIFY OBJECTS AND GROUP THEM INTO CLASSES AND SUPER CLASSES?

In OO, the analysis and design occurs in a top-down fashion, while construction is bottom-up process, because systems are built from pre-defined components that are assembled into a finished system. There are various views that lay emphasis on identifying object types based on their role they must perform, in relation to the concepts of objects in development. But the ones on the basis of which we designed objects in the real projects is summarized as:

- Interfaces and Presentations
  Objects that manages interaction with the user. Example – Menu/Query Screens etc.
- Control and Computation
  Objects that manage control in the application flow and decision making
- Data Management
  Objects that manage data structures, both internal and external to the application

We suggested a simple, four-step object-oriented system development life cycle-OOSDLC:

1. Design a model of the business activity
2. Construct classes to support the model
3. Assemble methods for the completed classes
4. Add interfaces to solve business problems

Bell Health care is one of the big specialty hospitals whose objective is to provide high quality health care while minimizing the cost to patients. The hospital is in process of developing an on-line processing system to handle tasks raging from patients registration to billing, which include inquiries, patient treatments and hospitalization, laboratory results, staff tracking and allocation. **We name the system as HMS – Hospital Management System.**

The first of the activity is to design a model of the business activity. After you have completed the process of system analysis and design, you arrive at defining the various sub-systems. This activity also includes review of the requirements and an analysis of these requirements to specify a high-level object model. These sub-systems define the major business activity, as in our case of HMS is shown in Figure 7.4(a). They are *Patient*

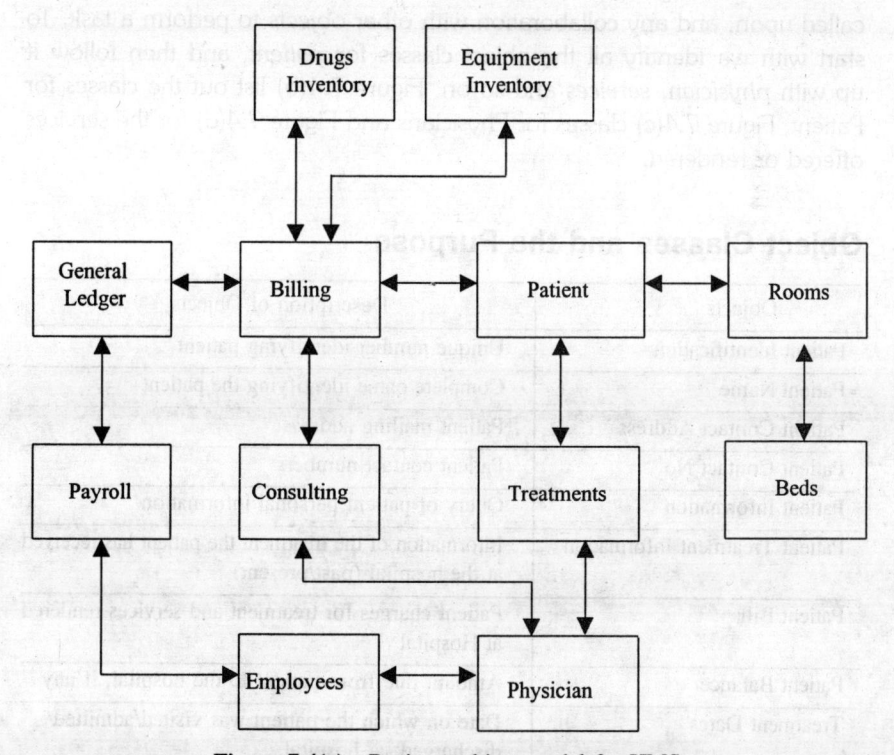

Figure 7.4(a) Business object model for HMS.

Registration, Patient Treatment, Patient Billing, Physician, Drugs and Services. What we do basically is to try and focus on **key abstractions.** Key abstractions are the ones which refer to entities that are at the heart of problem. These entities define the boundaries of the system. Eventually, these key abstraction become classes in the analysis model.

> **Note**
> In deriving key abstractions, it is very important to keep in mind that the level of abstractions should not be extreme, otherwise it will lead to a general definition of the entities in hand and will be confusing to work with objects.

Another key activity in OOA is to identify objects and group them into classes and then into super class. **This is the second activity in OOSDLC, i.e. Construct classes to support the model**

When designing systems using OO methodology, we model **real-world objects** with software objects. This way we identify relevant object classes from the application domain. In the HMS, we have identified real world objects which include **physical entities** like patients, physicians, rooms, as well as **concepts** like treatment, consulting and billing.

While defining the classes you must indicate the purpose, its task when called upon, and any collaboration with other objects to perform a task. To start with we identify all the object classes for *patient*, and then follow it up with *physician, services* and so on. Figure 7.4(b) list out the classes for Patient, Figure 7.4(c) classes for Physicians and Figure 7.4(d) for the services offered or rendered.

Object Classes and the Purpose

Objects	Description of Objects
Patient Identification	Unique number identifying patient
Patient Name	Complete name identifying the patient
Patient Contact Address	Patient mailing address
Patient Contact No	Patient contact numbers
Patient Information	Query of patient personal information
Patient Treatment Information	Information of the treatment the patient has received at the hospital (past/present)
Patient Bill	Patient charges for treatment and services rendered at Hospital
Patient Balance	Amount due from patient to the hospital, if any
Treatment Dates	Date on which the patient was visited/admitted/discharged at hospital

Figure 7.4(b) Table of object classes for patients.

Objects	Description of Objects
Physician Identification	Unique number identifying physician
Physician Name	Complete name identifying the physician
Physician Address	Physician address
Physician Contact No	Physician contact numbers
Physician Information	Query of the physician information
Physician Specialty	Area of specialization of the physician
Physician Services Rendered	Services attended to by physician

Figure 7.4(c) Table of object classes for physicians.

Objects	Description of Objects
Room Identification	Room types and characteristics
Beds Identification	Unique no identifying the beds in hospital
Ailment Identification	Identification of the type of ailment
Patient Service Charge	Default amount the patient is charged for the service
Patient Treatment Charge	Total amount charged to the patient for the treatment.
Physician Consulting Charge	Consulting fees charged by physician.
Room and Bed Charges	Per day charges for the rooms and bed
Room/Bed Occupancy	Room information bed wise
Main Menu	Entry Menu to HMS Present user with choices for navigation Determine what and when the users entered

Figure 7.4(d) Table of object classes for services.

Guidelines to Identify and describe Classes

When identifying classes, we must look for:

- Tangible things
  Example: hospital, person, drugs, rooms, beds etc.
- Roles played
  Example: patients, nurse, doctors, staff etc.
- Events that take place
  Example: consulting, treatment, operations, billing, payroll etc.

Every time a new class is to be created, existing classes must be thoroughly examined to determine whether the new class is required. In our example, we could have only one class as *employees*, instead of defining one for Physicians. But since each has a distinct role and characteristics, we separate them. Instead, the Physician class will include doctors, nurses, specialists etc.

When selecting and qualifying the class objects, one of the important points to be considered is that it relates the way business/operations are carried out in the system. Hence the object classes selected and defined are the ones whose business logic may bring about change in the information recorded from time to time. To illustrate this, take the object class **Patient contact address and Patient contact no.** Although in normal sense both are integral to contact information of the patient, but in object sense, either of mailing address or contact no can change independently. Hence both are treated as separate object class, as each object must be encapsulated. Also the name of a class should reflect its intrinsic nature and not just the role that it plays.

Further, we see that there are two characteristics of the patient we want to model: the things a patient **knows** and the things a patient **can do**. Thus in our case, a patient **knows** his or her name, address and contact numbers. While a patient **can do** includes *change of address* and *contact numbers*. In OO terms, the **things patient knows are called attributes and the things it can do are called methods.**

Next we group all the individual classes into super class as shown in table in Figure 7.4(e).

Super Classes	Object Classes	Super Classes	Object Classes
Identification	• Patient Identification • Physician Identification • Room Identification • Beds Identification • Ailment Identification	Name	• Patient Name • Physician Name
Display	• Patient Information • Patient Treatment Info. • Physician Information • Services Rendered by Physician • Room/Bed Occupancy • Main Menu	Charges	• Patient Treatment Charge • Patient Service Charges • Physician Consulting Charge • Room and Bed Charges
Address	• Patient Address • Physician Address	Contacts	• Patient Contact No • Physician No

Figure 7.4(e) Table showing objects and their super classes.

By splitting the reporting, user interface, and list services, you allow for future flexibility and insulate from the implementation of these services. For example, you might choose to implement the HMS system in any of the operating system, DOS, Windows, OS/2 or Unix. If you separate the library domain from the user interface, you are able to provide generic services to

Object-Oriented Analysis and Design **163**

the library, such as displaying screens, collecting and verifying inputs, etc. This also supports the concept of software layering and defining specific interfaces between subsystems.

The third of the activity is to assemble methods for the completed object classes. Here for each *objects you identify the services associated*. This phase seeks to identify and further define the classes, objects, attributes, and operations found in the prior phase. In some cases, this also would include specification of object behaviours. **It is important to remember that at this point, no attempt is made to specify the underlying details of each object in terms of its implementation.**

Table shown in Figure 7.4(f) describes the roles for each of the objects. Here you as a designer identify interactions between classes, or their collaborations. You might view this activity as defining the first part of a message by identifying the receiving object, without defining the detailed flow of control and data involved in message.

Object Class	Purpose/Description of Object Class
Patient Identification	Accept/Validate/Display unique identification
Patient Name	Accept/Display/Search Identify the patient
Patient Contact Address	Accept/Display Patient Mailing address
Patient Contact No	Accept/Display Patient contact numbers
Patient Information	Add/Edit/Delete/View patient personal information records
Patient Treatment Information	Add/Edit/Delete/View/Search treatment records of the patient received at the hospital. Validate/update ailment type Validate/update room/bed information of patient on daily basis during the time at hospital. Validate/update physician identification for the patient.
Patient Bill	Display/Print patient charges towards treatment and services at hospital.
Patient Balance	Compute/Accept/Display amount due from patient to the hospital
Treatment Dates	Accept/Validate/Display dates on which the patient was visited/admitted/discharged at hospital.
Physician Identification	Accept/Validate/Display unique identification
Physician Name	Accept/Display/Search the physician name
Physician Address	Accept/Display Physician address
Physician Contact no	Accept/Display Physician contact numbers
Physician Information	Add/Edit/Delete/View/Search physician information records

164 Workbook on Systems Analysis and Design

Object Class	Purpose/Description of Object Class
Physician Specialty	Add/Edit Area of specialization of the physician
	Query physician for particular specialization needs of a patient
Physician Services Rendered	Add/Edit/View/Query services attended to by physician at other hospitals
Room Identification	Add/Validate/Delete/Display room types and its characteristics
Beds Identification	Add/Validate/Delete/Display identification of beds in hospital
Ailment Identification	Add/Validate/Delete/Display identification of the ailment
Room and Bed Charges	Add/Edit/Delete/View per day charges for the rooms and bed at Hospital
Room/Bed Occupancy	Accept/Display accommodations details To provide for additional patient requirements
Patient Service Charge	Accept/Edit/Display the default amount the patient is charged for the service Validate/accept ailment types Validate/accept physician consulting charges
Patient Treatment Charge	Compute/Display/Edit total amount charged to the patient for the treatment.
Physician Consulting Charge	Consulting/Operations fees charged by physician.
Main Menu	Entry Menu to HMS Present user with choices for navigation Determine what and when the users entered

Figure 7.4(f) Table showing object and the methods, which act upon it.

The fourth and final activity is to add interfaces to solve business problems. In this activity the designer *establish interactions, services rendered and required by each object*. This phase seeks to define messages passed between objects and the services that they will provide to other objects. Object responsibilities and behaviours are defined in terms of interfaces. Here its interesting to note how this is similar to a client-server environment where one class (a client) requests some kind of action or dependency on another class (server).

Table shown in Figure 7.4(g) lists the objects and the message no as the identification for calling the objects. As shown in Figure 7.4(h), when you want to display patient information, you basically display the objects associated like Patient identification, Patient name, Mailing address, Contact no, Ailment type and Treatment dates. This has been illustrated within the brackets in the message column.

To make it more clear, from the table you can see that the class *Patient Identification* is required for the object classes Patient Name, Patient Information, Patient Treatment Information, Patient Bill, Patient Balance, Room/Bed occupancy, Patient Treatment charges.

Object-Oriented Analysis and Design 165

Object Class	Message	Purpose description of Object Class
Patient Identification	1 [2, 5, 6, 7, 8, 29, 31]	Accept/Validate/Display Unique identification
Patient Name	2 [5, 6, 7, 8, 29, 30, 31]	Accept/Display/Search Identify the patient
Patient Contact Address	3 [5]	Accept/Display Patient Mailing address
Patient Contact No	4 [5]	Accept/Display Patient contact numbers
Patient Information	5	Add/Edit/Delete/View patient personal information records
Patient Treatment Information	6	Add/Edit/Delete/View/Search treatment records of the patient received at the hospital. Validate/update ailment type Validate/update room/bed information of patient on daily basis during the time at hospital. Validate/update physician identification for the patient.
Patient Bill	7	Display/Print Patient charges towards treatment and services at hospital.
Patient Balance	8	Compute/Accept/Display amount due from patient to the hospital
Treatment Dates	9	Accept/Validate/Display dates on which the patient was visited/admitted/discharged at hospital.
Physician Identification	15 [16, 17, 18, 19, 20, 21]	Accept/Validate/Display Unique identification
Physician Name	16 [19, 20, 21]	Accept/Display/Search the physician name
Physician Address	17	Accept/Display Physician address
Physician Contact no	18	Accept/Display Physician contact numbers
Physician Information	19	Add/Edit/Delete/View/Search physician information records.
Physician Specialty	20	Add/Edit Area of specialization of the physician Query physician for particular specialization needs of a patient
Physician Services Rendered	21	Add/Edit/View/Query services attended to by physician at other hospitals
Room Identification	25 [26,28, 29]	Add/Validate/Delete/Display room types and its characteristics
Beds Identification	26 [28, 29]	Add/Validate/Delete/Display identification of beds in hospital
Ailment Identification	27	Add/Validate/Delete/Display identification of the ailment
Room and Bed charges	28	Add/Edit/Delete/View per day charges for the rooms and bed at Hospital
Room/Bed Occupancy	29	Accept/Display accommodations details To provide for additional patient requirements
Patient Service Charge	30	Accept/Edit/Display the default amount the patient is charged for the service Validate/accept ailment types Validate/accept physician consulting charges
Patient Treatment Charge	31	Compute/Display/Edit total amount charged to the patient for the treatment.
Physician Consulting Charge	32	Consulting/Operations fees charged by physician.
Main Menu	33	Entry Menu to HMS Present user with choices for navigation Determine what and when the users entered

Figure 7.4(g) Message request and its usage for other classes

166 *Workbook on Systems Analysis and Design*

```
                          ┌──── Patient Identification
                          │
                          ├──── Patient Name
                          │
┌─────────────────────┐   ├──── Mailing Address
│ Patient Information │───┤
└─────────────────────┘   ├──── Contact No
                          │
                          ├──── Ailment type
                          │
                          └──── Treatment Dates
```

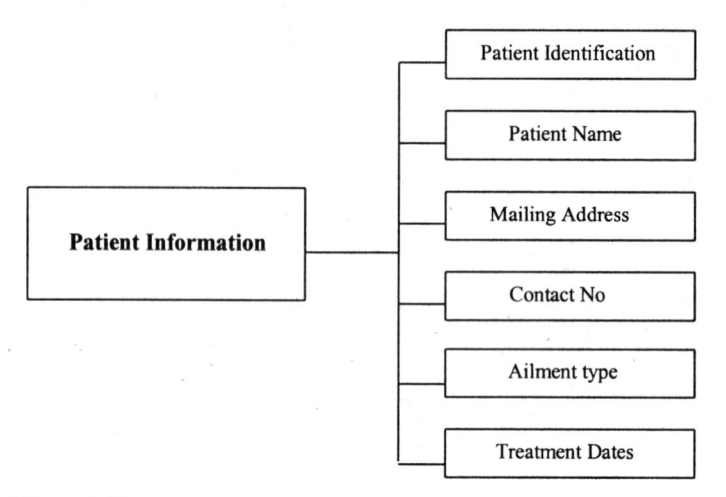

Figure 7.4(h) Patient Information and its dependencies on other objects.

Figure 7.4(i) shows how the main menu is structured. It has four subsystems, namely **Patient, Physician, Drugs and Consulting.** Within Patient subsystem, we have further 3 choices or subsystems. This way you do an efficient grouping based on classes for implementation purpose.

Once the above four activities mentioned are over, each object classes are designed and defined to detail level. The construction and implementation parts of these objects start now. Here the client/server and contractual relationships are defined, i.e. the message passing among clients objects and server objects is controlled by developing the functions as shown in Figure 7.4(j). **The table contains the functions for the object classes. Message passing among client objects and server objects is controlled by these functions. Within parenthesis it specifies the type of parameter that these objects can receive and send.**

```
                    ┌── Patient ────┬── Registration
                    │               │
                    │               ├── Treatment
┌───────────┐       ├── Physician ──┤
│ Main Menu │───────┤               └── Billings
└───────────┘       │
                    ├── Drugs
                    │
                    └── Consulting
```

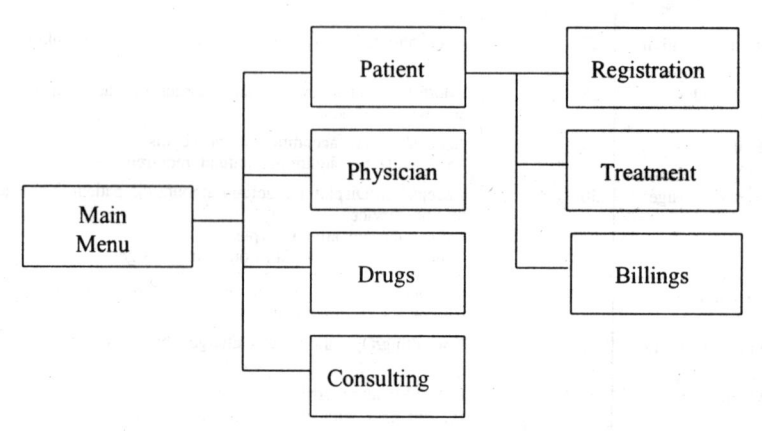

Figure 7.4(i) Structure of main menu.

Object-Oriented Analysis and Design 167

Object Class	Signatures
Patient Identification	Get_Patient_Id(Integer) Validate_Patient_ID(integer) returns (Boolean)
Patient Name	Get_Patient_Name(Text)
Patient Information	Add_Patient_info(Boolean) Modify_Patient_info(Boolean) View_Patient_info(Boolean) Delete_Patient_info(Boolean) Search_Patient_info(Boolean)
Patient Treatment Information	Add_Patient_treat_info(Boolean) Modify_Patient_treat_info(Boolean) View_Patient_treat_info(Boolean) Delete_Patient_treat_info(Boolean) Search_Patient_treat_info(Boolean) Validate_room_bed_no(integer) returns (Boolean)
Patient Bill	Compute_Patient_Bill(Boolean) Modify_Patient_Bill(Boolean) View_Patient_Bill(Boolean) Delete_Patient_Bill(Boolean) Validate_room_bed_no(integer) returns (Boolean)
Patient Balance	Get_Patient_Balance(Amount)
Physician Identification	Get_Physician_Id(Integer) Validate_Physician_ID(integer) returns (Boolean)
Physician Name	Get_Physician_Name(Text)
Physician Information	Add_Physician_info(Boolean) Modify_Physician_info(Boolean) View_Physician_info(Boolean) Delete_Physician_info(Boolean) Search_Physician_info(Boolean)
Physician Specialty	Get_Physician_speciality(Text) Search_Physician_Info(Text) returns Physician Information view
Physician Services Rendered	Display_service(Text) Add_Physician_service(Boolean) View_Physician_service(Boolean) Edit_Physician_service(Boolean) Search_Physician_service(Boolean) Search_Physician_speciality(Text) returns Physician Name
Room/Bed Occupancy	Add_Room_Bed_allocation(Boolean) Modify_Room_Bed_allocation(Boolean) Delete_Room_Bed_allocation(Boolean) Display_Room_Bed_allocation(Boolean) Query_Room_Bed_allocation(Boolean) Room_Bed_occupancy_period(Boolean)
Room/Bed Charges	Add_Room_Bed_charges(Boolean) Modify_Room_Bed_charges(Boolean) Delete_Room_Bed_charges(Boolean) Query_Room_Bed_charges(Text) returns Room/Bed allocation Room_Bed_charges(Boolean)

Figure 7.4(j) Object classes and its implementation functions.

168 Workbook on Systems Analysis and Design

In this way you go about constructing the functions along with the parameter these objects receive or send. Thus we see that object-oriented, programming is a process of assembling objects to support the functionality for a system. Unlike traditional procedural language, objects act in an asynchronous manner, and can only interact via the passage of messages.

Developing Class Diagrams

The next step is to establish class diagrams. The class diagrams record information about a class, which are:

- Name of the class
- Services provided by the class
- List of class attributes

By creating class diagrams you communicate information about a system in an easy-to-understand, visual format. Figure 7.4(k) illustrates a class diagram. In this you could have included type information by prefixing each of the attributes with its data type.

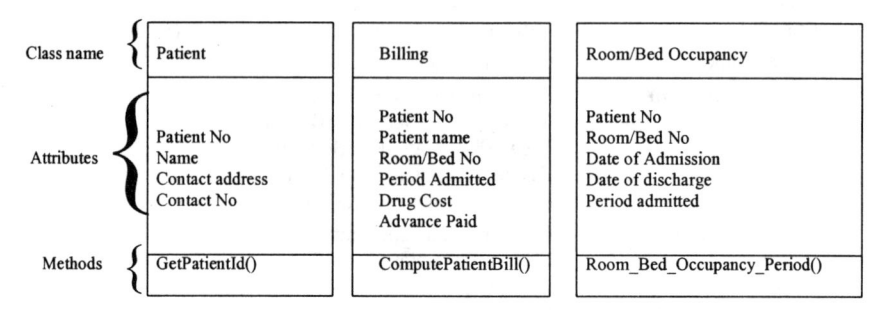

Class name	Patient	Billing	Room/Bed Occupancy
Attributes	Patient No Name Contact address Contact No	Patient No Patient name Room/Bed No Period Admitted Drug Cost Advance Paid	Patient No Room/Bed No Date of Admission Date of discharge Period admitted
Methods	GetPatientId()	ComputePatientBill()	Room_Bed_Occupancy_Period()

Figure 7.4(k) Class diagram.

Classes in OO are often associated with one another. A class diagram can include multiple classes and are depicted by associations as shown in Figure 7.4(l).

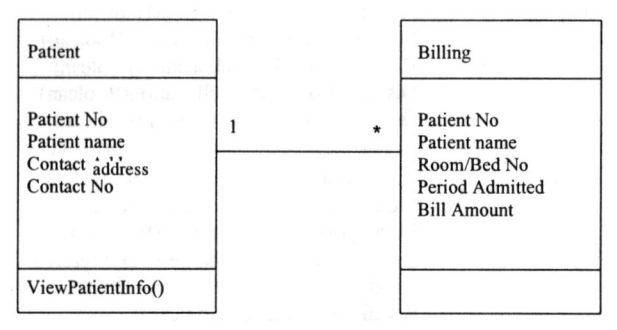

Patient		Billing
Patient No Patient name Contact address Contact No	1 *	Patient No Patient name Room/Bed No Period Admitted Bill Amount
ViewPatientInfo()		

Figure 7.4(l) Class diagram indicating associations.

The "1" indicates that each *Billing* is associated with exactly one *Patient*, whereas the "*" indicates that a *Patient* can be associated with zero, one or more *Billing*. These relationships are on the lines of entity-relationship diagrams used to model database relationships. In OO also the possible associations between classes are:

- **One to One links**
- **One to Many links**
- **Many to Many links**

The similarity of relationships between ER diagrams and class diagrams is only in its depiction. While defining association in OO Classes encapsulate data as well as behaviour, but in ER its only establishing database tables.

This way, we construct class diagrams establishing associations and a final relationship diagram is depicted as shown in Figures 7.4(m). In the association between *patient, rooms* and *beds, one room* can have *one or more* beds, *one or more patients* can share *one room*, whereas *one patient is* attached to *only one bed*.

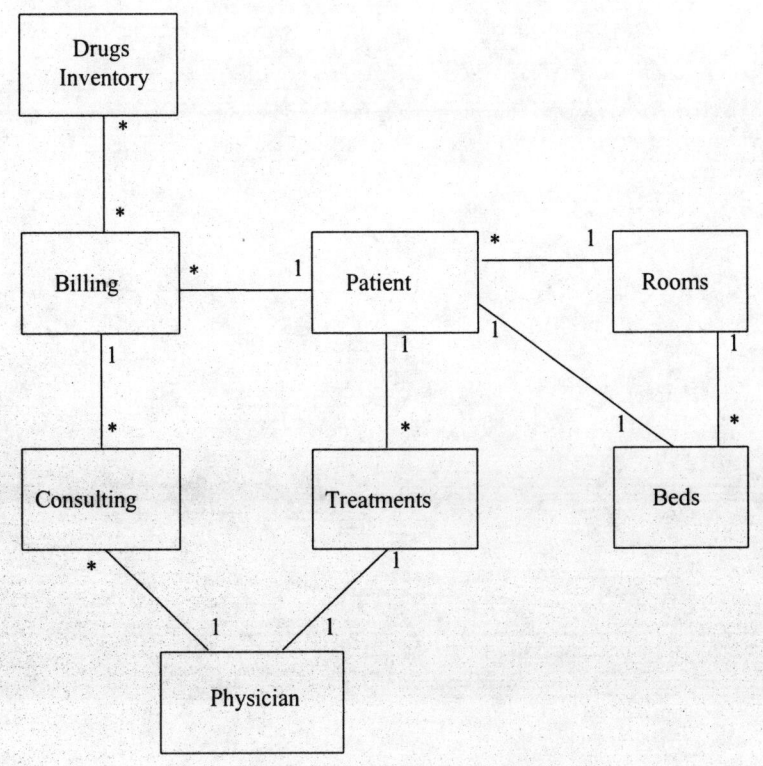

Figure 7.4(m) Final object relationship model.

170 Workbook on Systems Analysis and Design

This brings you to the end of object-orientation design. What we have finally in hand is:

- Business Objects (Processes and Entities of Level-1 DFD in SSAD)
- Classes (Modules in Level-0 in SSAD)
- Object relationship model (ER diagram in SSAD)
- Attributes (Data elements in SSAD)
- Methods (Sub processes as per Level-2 in DFD)

Based on these inputs, the process of data collation and normalization starts, which is same as per SSAD. All the attributes based on class diagrams are put into a normalized form and are then worked out as 1NF, 2NF and 3NF. Once the data modeling is completed, process specifications for the objects and the methods which act on it is worked out. For these the SSAD process specification tolls like structure charts, decision tree, decision tables and structured English are followed.

Practice for Chapter 7

#1

Property Management Agency (PMA) is a property-management company that oversees commercial rental properties. As part of its rent-collection operation, PMA maintains data about the properties and dwellings/units it manages. For each unit, PMA tracks the unit address, owner name, size of the place (in both square feet and square metre), number of rooms, and special amenities (e.g. private restrooms, private entrance, reception area, balcony, bath tubs, etc.).

These units are leased by tenants who sign a lease agreement stating the beginning occupancy date, ending occupancy date, lease terms, deposit amount paid, monthly rent, and rent-due date. PMA also maintains data about each tenant, including tenant name, tenant office phone, nature of business, and at least two references. Based on the tenant profession, PMA allocates a rating to him. Each lease agreement represents only one tenant and one unit. However, each tenant may sign many leases (e.g. renew the current lease or lease another unit), and each unit may, over time, be covered by many lease agreements.

Based on the above do the following:

(a) Design a model of the business activity.

(b) Construct classes to support the model

(c) Create an object relationship model for Property Management Agency (PMA).

#2

Data about employees is maintained on the three source documents shown in three documents below. Employee fills the first of the document every month for authorized deductions. The second document is time sheets for each employee as generated by recording system. The third document is the time sheet generated after payroll execution.

Employee Payroll Information
Applicant: Please provide the following information for purposes
Full Name: Hema Raman   Employee No: 121   Tax Status: Professional   Authorized Deductions: Exemptions: 1   Type   Amount   Month: 03/99   1. PF   200/-   2. Parking   45/-
For Office use only
Department: Marketing   Days Present: 25   Days Absent: 5

Timecard for:	Hema Raman
Employee No:	121
For Month:	03/99
Date:	1/03/99 - 26/03/99

Date	Time -In	Time-Out
1/03	9.36	5.50
2/03	9.40	6.30
3/03	9.42	4.55
Etc.		

Time Sheet for SAP Department
Pay Period Ending 26/3/99

Emp.No.	Name	Gross Pay	Deduction	Net Pay
121	Hema Raman	6000.00	235.00	5765.00
122	Rajshekar	9000.00	575.00	8425.00
123	Sanjay Saxena	9000.00	500.00	8500.00
etc.				

Based on the above documents, create an object relationship model-to-model the relationships between classes. Also identify attributes for the object classes.

#3
Jet airlines offer numerous flights between various destinations across India every day. Information is maintained about flights, which are flight number, departure time and city, destination city and arrival time, class and seats available. To reserve a seat on any of the flights, a customer calls a toll-free number and gives the ticketing agent his/her details which are: name, address, telephone numbers and credit card number. If the seat is available the ticketing agent issues a ticket stating the ticket number, flight details (as mentioned above), ticketing conditions and fare. Although each flight can be booked by many customers and each customer can book in many flights, Jet ticketing policies requires that each customer must have a ticket and boarding pass to board each flight. The boarding pass is issued at airport departure lounge in exchange for the ticket issued.

Create an object relationship model of Jet Airlines ticketing system. Show attributes and identifiers for object classes in your model.

#4
The SAREGAMA agency run by Fiddle Maestro, a musician agent who organizes music concerts across the country. Fiddle has entered into contracts with troupes, bands, individual musicians and clubs for arranging and

conducting concerts. His income is through a fixed percentage of the proceeds in payment. Fiddle gathers and maintains information about clubs, bands and musicians in the information booklet. In another folder she maintains the booking contracts. Shown below is the sample of the data stores of the information stored by Fiddle.

Sample Contract

Band:	W2K
Booking Date:	Feb 10, 1999
Restraints:	Full A/c
	No smoking
Booking Rate:	Rs, 25,000/-
Payment Terms:	Check for complete amount made from the order of SAREGAMA. Payment due within 7 days after booking date.

Sample Band/Musician card

Band Name:	New Millennium	Band:	W2K
Band Members:	Ray, Daisy, Shankar Udit, Babla	Phone:	91-22-5831258
Music:	Rock		
Features:	Lots of indian tunes Great orchestra 10,000 watts music	Minimum Rate:	Rs. 25,000/- per night
Agency Fee:	15 % of Gate fees		

Sample Club card

Club Name:	Wilmington	Club:	WIL
Address:	Golf Drive, Powai Mumbai 50	Phone:	91-22-5831258
Manager:	Vikram Garg		
Features:	Full A/c Seat 3000 people	Rate:	Rs. 15,000/- per night
Play Hours:	5 pm - 11.30 pm		
Cancellation Policy:	72 hours notice, Refund 85 % of booking amount		

Based on the documents, create an object relationship model with attributes and identifiers to model the object classes and relationships in Fiddle contract-management system.

8

Question Bank

EXERCISES

#1
The administrative staff of the Executive Director office (EDO) of COSL wants a system to create special reports from the payroll data of accounts department. Presently the EDO office produces couple of reports each month for the Director. The EDO staff would like to use a computer system that would create those reports every month. In addition to preparing these reports, the EDO staff compiles and distributes reports of payroll to all the companys staff, given by accounts department.

    (a) Who are the users of the proposed system?

    (b) Draw a consolidated context diagram for the proposed system.

#2
Reshmi Singh is responsible for preparing a budget for advertisements issued by your company for placement opportunities. She has figures for advertising costs in various newspapers and magazines on different pages and locations. The advertisements are released based on the request from different departments. Each department has an allocated yearly budget for their advertisement needs. The bills sent by the media are scrutinized by her and sent to accounts department for settlement.

    However, because of the number of advertisements, media types and locations, she finds it difficult to do so without assistance. She has suggested a computer-based decision support system and you have been asked to study and design the system.

    (a) Make a list of the external entities which send to or receive information from the agency.

    (b) For the Prime Placements, make a list of the documents and information flowing from and to the system.

    (c) Prepare a user-level context diagram that summarizes the information of Reshmi department.

#3

Marketing department of Cybex System Europe (CSE) is into software consulting business. One of their marketing arms is into scouting for opportunities to place consultants on projects with the client. The system logs information of opportunities worldwide. For each of the opportunities the marketing department forwards consultants from the database, based on skill sets, relevant experience, functional and educational background required by client.

The system must capture consultant information that is short-listed by CSE. These information are consultant's personal detail, his academic background, work experience and skill sets. The system will also have the database of various clients/companies. The client screens and shortlists the consultants forwarded for interview. Once selected, the consultants are placed with the client and are billed for the services based on the agreed terms and rates. When a consultant is back from the project, he is made available for future opportunities.

(a) **Identify all the processes, data flows and data stores.**

(b) **Draw a consolidated context diagram for the proposed system.**

(c) **Draw all the levels of DFD to explain the above process and data flows to lowest level.**

(d) **Based on the DFD output, list out the entity relationships between data stores and normalize the data structures.**

#4

Given below is a scope and overview of project billing system for COSL. Overseas employees of COSL send in timesheets of their working hours at the client's place on regular basis. These time sheets are entered into the system which calculates the number of billable hours of these employees and the revenue earned by each employee for the company. The system generates and prints invoice to be sent to client. When the remittance comes from the client, these are marked against the outstanding invoices.

The following are the final data stores derived from data flow diagram.

- Client Master
  Input: Client details
  Process: Create client codes and maintain information pertaining to clients.
- Project Master
  Input: Project details
  Process: Maintaining project codes, project period and project logistics.
- Employee Project Details
  Input: Details of employee with respect to his project.
  Process: To keep track of employee rates and billing details

- Employee Timesheets Entry
  Input: Timesheets sent in by the employees
  Process: To record the employee daily timesheets (hours worked, overtime hours)
- Billing Details
  Input: Employee timesheets
  Process: Calculate Employee billed amount
- Invoice Details
  Input: Billing amounts from billed table
  Process: Generate and print invoices
- Remittance Details
  Input: Invoice details
  Process: Update remittance details

(a) **Draw a data flow diagram to cover the above mentioned activities.**

(b) **Identify all the data elements required to create a data dictionary.**

(c) **Draw a functional dependency diagram for the identified data elements.**

(d) **Normalize the functional dependency diagram to 3NF by process of normalization.**

(e) **Validate your results by means of an E-R diagram.**

#5
Develop E-R diagrams (designating the type of relationship) for the following descriptions:
- (a) Customers place orders
- (b) Deliveries of parts are made to Customers
- (c) Vehicles are owned by persons
- (d) Athletes take part in sporting events
- (e) Customer book rail tickets
- (f) Students enrolling for courses.

#6
Draw E-R diagrams showing the cardinality for the following descriptions:
- (a) Students select subjects. Many students can take each subject, and each student can take as many subjects.
- (b) Persons apply for loans. Each loan must be made to one person, but each person can submit many applications.
- (c) Various garments are produced. Each garment is made up of various materials.

(d) An operator can work on many machines, and each machine has operators. Each machine belongs to one department, but a department can have many machines.

(e) An invoice is sent to one customer, and there can be many invoices sent to the same customer.

#7
Draw E-R diagram with attributes, cardinality and identifiers for the following:

(a) Departments are identified by DEPT-NO and have a budget allocated for them. A department can handle and manage many projects, but only one department manages each project. Projects are identified by PROJECT-NO and have a START-DATE.

(b) An order is prepared for each item. The order with a unique ORDER-NO and ORDER-DATE is made for any number of parts (identified by ITEM-NO) and QUANTITY-ORDERED. Each order is made to one supplier. Supplier has a unique SUPPLIER-CODE, NAME and ADDRESS.

(c) A fault occurs on one item of equipment. A log-book contains FAULT-NO, FAULT-DATE and FAULT-DESCRIPTION. Each item of equipment and TYPE. Each such item is located in one building which has a unique BUILDING-NAME and ADDRESS.

#8
Below are the statements about order processing in an organization. You are required to construct an E-R diagram from these statements.

(a) Each employee has a department to which he belongs and is identified by EMPLOYEE-NO, and has EMP-NAME, SEX and DATE-OF-BIRTH.

(b) Employees place orders, which are identified by an ORDER-NO and have an ORDER-DATE, DESCRIPTION, ITEMS-CODE, ORDER-AMOUNT. Only one person is responsible for a given order, but a person may be responsible for many orders.

(c) The organization manufactures some of the items and the rest are procured from different vendors. The employee, responsible for the order, makes formal request, which is identified by a REQUEST-NO. They nominate a START-DATE and END-DATE for each request.

(d) A number of jobs can be created by a section in response to a request. Each job is identified by a JOB-No and has a COST. All jobs for one request go to the same section, which is identified by SECTION-ID and has one employee a MANAGER heading it.

(e) Each job uses a QTY-USED of one or more materials. Materials are identified by MATERIAL-ID and have MATERIAL-DESC.

#9

An art exhibit company exhibits various sculpture and paintings of different artists regularly at different art galleries across the country. Each piece of painting has a PAINTING-CODE, PAINTER-NAME and EVALUATED-VALUE. Similarly each sculpture has a SCULPTURE-CODE, SCULPTURE-NAME and APPROXIMATE-VALUE and AGE.

Paintings and sculptures are exhibited at different galleries and carry the LOCATION-CODE and GALLERY-CODE to identify the place and location of the exhibit. Each art object appears in one gallery only. The DATE-PLACED-IN-GALLERY is kept for both objects. Note that PAINTING-CODE and SCULPTURE-CODE are unique for each type of objects with a PAINTER-NAME or a SCULPTURE-NAME.

Develop a decision tree and a decision table for the above:

#10

An organization services a variety of equipments. Following a fault report, which may be through letter or phone, a fault report form is filled in and sent to the despatch centre. A mechanic is selected at the despatch centre, and a partial report is prepared. This report includes the MECHANIC NAME, the equipment to be serviced, the FAULT-DESCRIPTION and the expected DATE and TIME of REPAIR. After the mechanic services the repair, details of the work are entered on the repair report. This includes the actual date and time spent, the parts used to service the equipment. After the repair form is received, the costing department works out the cost for the parts used, based on the equipment warranty. The invoice is sent to the customer, who settles the bill with the settlement department.

Draw an organization level context diagram for the proposed systems

#11

A theatre accepts telephone and fax bookings for its forthcoming performances up to one month in advance. The bookings are in the form of personal, corporate and agency bookings.

When a member of the public makes a booking by telephone, either a credit card number is taken and the booking is firmly made, or the reservation for the seat is taken. In the latter case, the public is given the option to either collect the ticket by paying the required amount, or the ticket can be delivered at his place, by collecting a nominal delivery charge in addition to the ticket amount. In the same way, it works for fax message.

In the case of corporate booking, on receipt of block-booking request by phone or fax, the theatre books the required seats, and sends the tickets along with the invoice. The invoice must be paid by the corporate at least two days before the show date, failing which the tickets stand cancelled, and the corporate is informed about the same.

In the case of an agency, on receipt of block-booking request by phone or fax, the theatre runs a credit check on the agency account and makes a confirmation note for requested seats that are available. This is sent to the agency as a confirmation of booking as well as an invoice. At the end of each month the theatre sends a statement of account to each agency. Each agency is allocated a fixed quota of seats, which they can book.

Half an hour before the performance starts, all those seats that are reserved by the public, but the tickets are not collected, are released for general sale. The theatre also receives inquiries on seat availability from general public, corporate and agency.

(a) **Draw an organization level context diagram for the proposed system.**

(b) **Draw Level-0, Level-1 and Level-2 DFD starting from level-0 till illustrating the data flows, processes and data stores to carry out the inquiry, booking, cancellation, refunds and invoicing for the Theatre.**

(c) **Normalize the data stores by process of normalization.**

(d) **List out the activity and design the menu and user interface for the system.**

#12

The following is the manual operations at the head office of Spencer Supermarket. The store wants to automate their back office functions at the head office.

Each branch of Spencer submits a daily cash report plus supporting documents to the head office by courier. The area manager's office receives these reports, and processes them. The staff at area manager's office carries out checks on the arithmetic of the cash reports and on submitted bank deposit slips and petty cash vouchers. All these documents are then passed to the cashier.

Each day's cash report is summarized and entered into a cash analysis book in the cashier's department. This cash analysis book forms part of the main cashbook into which weekly totals are entered. At the end of each week the cashier's department reconciles the cashbook with the bank pass book. The cash reports are then sent to the accounts department.

Every week each branch of Spencer sends the deliveries made by suppliers together with other stock movement details. These are sent to the area manager's office where the unit costs and sales prices are entered on the document and delivery records. The complete set of documents is passed to accounts department. The area manager office also receives delivery sheets sent from the company's warehouse. These are goods supplied to the branches. The cost and selling prices of these sheets are submitted to the accounts department.

The accounts department receives the stock movement forms, the direct and internal delivery sheets and the cash reports. The department then prepares a monthly report for each branch of Spencer for cost and selling price.

(a) **Identify all the processes, data flows and data stores.**
(b) **Draw a consolidated context diagram for the proposed system.**
(c) **Draw all the levels of DFD to explain the above process and data flows to lowest level.**
(d) **Based on the Level-2 DFD, list out the data entities and establish entity relationships by way of E-R diagrams.**
(e) **Identify the data elements part of data stores and normalize them to 3 NF.**

#13
Draw first level data flow diagram for the system consisting of the processes, mentioned below for doorstep household service.

Customer Requirements. Customers sends the list of items, along with the quantity to the organization.

Buying Requirements. The organization buys the items asked by the customers, packs as per the quantity required and gets them bundled for despatch. Before the items are bundled, the list is checked for any missing items.

Item Delivery. During the delivery of items to customers, an invoice is prepared for the cost of the items. Based on the amount of goods ordered, a service charge is added to the invoice.

Receive Payments. A service is completed when payment is received from the customer. It can be through cash or cheque.

#14
Draw a functional dependency diagram showing functional dependencies between the capitalized attributes in the following problem:

Policies identified by POLICY-NO can be established in an organization. Each policy has one DATE-SET-UP and is set up for one customer. Each customer has a CUSTOMER-ID. The customer also has an ADDRESS, but addresses can change. The START-DATE for the policy is kept for each customer. There is one RISK-LOCATION for each policy.

A policy can include any number of special items. Each special item has a unique SPECIAL-ITEM-NAME within the policy. The VALUE of each special item is recorded.

Claims can be made against policies. Each claim has a unique CLAIM-NO and is made on a given CLAIM-DATE and a CLAIM-AMOUNT. Any special items included in the claim are recorded, together with the ADDITIONAL-AMOUNT-CLAIMED for each special item.

#15

The gatekeeper at ESSEL WORLD amusement park, is given the following instructions for admitting the visitors to the park:

(i) Children below 3 feet height are admitted free.

(ii) Persons above 3 feet and less than 4 feet are charged half of the admission fee.

(iii) Person above 4 feet is considered an adult and is charged full admission fees. If an adult is accompanied with more than two children, the three children are admitted free.

(iv) If a person is above 4 feet and is a student, and shows his/her identity card, then the person is charged half the admission fee.

Develop a decision tree and a decision table for the above conditions for visitors.

#16

The following information of the data stores and data elements has been gathered from DFD. **Draw an E-R diagram for it.**

(i) Workers (identified by WORK-NO), work on machines (identified by MACHINE-NO) to knit and produce garments.

(ii) Various types of Garments (GARMENT-TYPES) are produced in the factory. Each GARMENT-TYPE has a description (GARMENT-DESCRIPTION) and is made up of different yarns (YARN-ID) and has a fixed composition of yarn (YARN-PER CENT) and quantity (YARN-QUANTITY).

(iii) The production of each garment is recorded as a job identified by JOB-NO. Each JOB-NO has a START-TIME and END-TIME and is performed by one worker on one machine. A number of garments of different kinds are produced on one job.

(iv) Other information required is:
- NAME and JOB of workers
- DATE-PURCHASED and NEXT-SERVICE-DATE for machines
- TIME-SPENT by each worker on a job
- NO-OF-GARMENTS produced on one job.

#17

Use structured English to describe the following system.

The system receives a batch of transactions, each of which has a key value. The system examines each transaction in the batch and, depending on the type, does the following:

(i) For transaction type TC (credits), the system stores records of type TC, but only if there is a type TD (debits) record for the same key.

Otherwise the system writes an error output. If there is a type MR record with the same key value as the type TC transaction, the type MR is converted to a type MT record.

(ii) For type TD transactions, the system stores the transaction as a record as type NR (not reconciled) in the main file.

(iii) For type MR transactions, the system stores the transaction type MR record, only if there is a pair of type TD and type TC records with same key values as the type MR transaction.

The system ensures that no two records of the same key value are present in the batch data.

#18

An invoice clerk receives invoices from suppliers. Each invoice contains information on:

(i) An order and supplier number
(ii) Items, Price and quantity of each item delivered
(iii) Total invoice amount.

The invoice clerk examines the invoice and compares it with both the order and a stock report. The stock report contains data on the goods received in the organization store from various suppliers. This data includes the order number and the supplier who delivered the items.

If the items on the order, the invoice and the stock report match, then the invoice clerk checks the total invoice amount. If the amount is correct, the invoice clerk sends an authority to the accounts department to issue a cheque for the invoice. If the amount is incorrect, the invoice clerk adjusts the invoice and authorizes the accounts department to issue a cheque for the adjusted amount. At the same time, the invoice clerk prepares and despatches a vendor memo advising of the adjustment.

If the item on the invoice does not match the stock report, but do match the order, the invoice clerk first checks for the correctness of the report. If it is correct, the clerk makes an adjustment to the invoice amount, authorizes accounts department to prepare a cheque for the adjusted amount and prepare a vendor memo advising the adjustment.

A stock memo is sent intimating about further items to be received against the order, and issuing a supplementary order number to both the store and the supplier.

Prepare a decision table to illustrate the activities of the invoice clerk as described above.

#19

The daily invoicing routine for an invoice clerk is as follows. At the beginning of the day the stock of sales order with the corresponding despatch notes is collected and processed. The details of each order are checked against the details of the relevant customer account, and a discount is calculated for each customer. The customers are allocated a fixed credit-limit within which they can get credits.

A customer is allowed 5% off the list price. There is also a special 5% discount for any customer who has been ordering regularly for the past 5 years, provided his order is more than 1 million Rs. Otherwise, any order over 1 million Rs is given a bulk discount of 7% of the list price. If the total invoiced amount exceeds customer credit limit, an approval from the manager is taken before the despatch of the invoice. When all the invoices are despatched a note of the total invoiced amount is sent to the manager.

Represent the above by using structured English.

#20
A stationery supply company has a number of goods depots located around the country. The company's products are promoted by a number of salesmen who visit the current clients and prospective clients. The salesmen take orders from the clients and send these to the company depot, which is nearest to the client.

Draw an E-R diagram after identifying the entities and the relationships encountered in such a system.

#21
A car hire company has 50 cars at its disposal. Each car is identified by its registration number. These cars are based at one of seven garages, each of which is known by its name: Leuven, Tinen, Gent, and so on. Any car can be serviced at one of the three service centres. The service centres are at Brussels, Antwerp, Brugge. Any of the company's clients may hire an available car for one or more hours, collecting the car from the garage and leaving it at any other garage.

Draw an E-R Diagram after identifying the relationships encountered in such a system.

#22
You are designing a system, which will investigate the use that customers make of a new "Shop Till You Drop" supermarket. The management of the supermarket is interested in the following aspects:

(i) How large is the customer's family?
(ii) How far does the customer travel to reach the supermarket?
(iii) How often does the customer visit the supermarket?
(iv) How much do they spend per visit?
(v) What other local stores does the customer visit?

(vi) Is there any special product, which the customer buys from the supermarket?

(vii) Are the shopping hours convenient?

(viii) Are there any improvements, which the customer would like to see at the supermarket?

(ix) What newspapers/magazines does the customer read.

(a) Design a form, which the customer desk can use to collect this information.

(b) Design an input screen, which could be used to input the information collected on the above form.

#23

An organization is in the process of upgrading microcomputer hardware and software for all employees. Hardware will be allocated to each employee in one of the three packages.

The first hardware package includes a standard microcomputer with a colour monitor of moderate resolution and moderate storage capabilities. The second package includes a high-end microcomputer with high-resolution colour monitor and a great deal of RAM and ROM. The third package is a high-end notebook-sized microcomputer. Each computer comes with a network interface card so that it can be connected to the network for printing and e-mail. The note-book computers come with a modem for the same purpose.

All new and existing employees will be evaluated in terms of their computing needs (e.g. the types of tasks they perform, how much and in what ways they can use the computer). Light users receive the first hardware package. Heavy users receive the second package. Some moderate users will receive the first package and some will receive the second package, depending on their needs. Any employee who is deemed to be primarily mobile (e.g. most of the sales force) will receive the third package. Each employee will also be considered for additional hardware. For example, those who need scanners will receive them and those needing their own printers will receive them. A determination will be made regarding whether or not the user receives a colour or black and white hand scanner, and whether or not they receive a slow or fast, and colour or black and white printer.

In addition, each employee will receive a suite of software, including a word processor, spreadsheet and presentation maker. All employees will be evaluated for their additional software needs. Depending on their needs, some will receive a desktop publishing package, some will receive a database management system (and some will also receive a developer's kit for the DBMS), and some will receive a programming language.

Every eighteen months those employees, with high-end systems will receive new hardware and then their old systems will be passed on to those who previously had the standard systems. All those employees with the portable systems will receive new notebook computers.

Present the logic of this business process using structured English first, and then using a decision table and after that a decision-tree.

#24

Upon receiving an approved project request form, a data entry user uses a data entry program to establish the project information in the database. The approved project request form contains the following information, in the order of its appearance in the form.

PROJECT ID, PROJECT NAME, CLIENT ID (the identifier for the owner of the project), EMPLOYEE ID (Manager handling the project), PROJECT START DATE, ESTIMATED PROJECT TIME and PROJECT APPROX. COST.

Create a Input data entry screen for the above information.

#25

Shown below is record layout of project details: List any assumptions that you make to complete the information.

Prj. ID	Prj. Date	Client ID	Client Name	Address	Emp ID	Emp Name	Dept ID	Dept Name
01	07-01-97	103	CTE	12, Park	125	Srinivasan	D01	Project
01	15-05-97	145	CIC	G3Midas	134	Venkat	D02	SAP
01	17-09-97	75	BAPIS	13 Prince	111	Lata	D03	EDO
01	27-02-98	103	CTE	12, Park	125	Reshmi	D03	EDO

(a) **Normalize the record into first normal form**
(b) **Normalize the record into second normal form**
(c) **Normalize the record into third normal form**

#26

Consider the following two relations:

(i) EMPLOYEE (No, NAME, SALARY).
Is the employee number (No) functionally dependent on either NAME or SALARY, or on both?

(ii) TRANSACTION (TRAN-DATE, TRAN-TYPE, TRAN-NO, TRAN-AC, TRAN-AMOUNT).
(a) **Is the transaction no (TRAN-NO) functionally dependent on transaction account (TRAN-AC).**
(b) **In the above relation which combination of keys will identify a given transaction for an account.**

#27

Given below is a structure of existing database used to record consumer loan by Alpic Finance. In the current database one can store maximum of three loans per customer, including the recent loan availed. Now the company wants to expand the historical capabilities so that complete loan record are available for each consumer.

The attributes associated are:

NAME	Name of consumer
CNUM	Country citizen number of consumer
ADDRESS	Location where consumer resides
CITY	City where consumer resides
STATE	State where consumer resides
PIN	PIN code of the city
AMT1	Amount of most recent loan
DATE1	Date of most recent loan
RATE1	Payment rating for most recent loan
AMT2	Amount of second most recent loan
DATE2	Date of second most recent loan
RATE2	Payment rating for second most recent loan
AMT3	Amount of third most recent loan
DATE3	Date of third most recent loan
RATE3	Payment rating for third most recent loan

Construct the above relation and normalize to 3NF.

#28

Consider the following six data stores: STUDENT, COURSE, INSTRUCTOR, HOUR, CLASSROOM and GRADE established from DFD, in training requirements database.

Attributes identified in the above data stores are:

EMPNAME	Instructor name
EMPNO	Instructor identification no
EMPSUP	Instructor supervisor identification no
CLNAME	Training class name
CLNO	Training class number
CLDESC	Description of class
DTSCH	Date an instructor is scheduled to take class
DTPER	Period for which the instructor continues the class
DTREP	Date an instructor began reporting to a supervisor
STUDENT	Students associated with the class
HOURS	Duration required to complete the class
CREDIT	Number of credits associated with the course

Establish the relationship between them by way of functional dependencies and normalize the database into 3NF.

#29

Described below is a case of Manchester City Council Rent and Community Charge (MCRC) Centre, collected by Manchester Municipal Council (MMC).

The council is a large urban/semi-rural council that collects rents and charges from householders, tenants, and business people. Some 5 million people live within MMC limits. About one-third pay rents of some sorts, while another half pay water charges, street tax and community charges to the council. Most of these charges are collected by MCRC. Traditionally the city is split into five wards, and the rural parts into three areas. MCRC has offices in all these eight wards/areas. The MMC wants the MCRC operations to be computerized.

At present the collectors from each of MCRC go to collect rent and charges from tenants. Tenants can also go to these MCRC offices or the central office and pay directly. These payments are made on monthly basis and can be paid either by Cash/Cheque/Bank draft. The amounts collected at MCRC regional offices are collected and sent to the central office for processing and records are updated accordingly. The community charges are normally assessed once in a year and collected quarterly from each payer.

If a tenant does not pay any of these charges as scheduled, the MCRC office issues a warning note. If no payment or explanation is forthcoming within 10 working days, severe warning letter follows this. If again this is ignored after 20 working days, a solicitor's notice is sent and proceedings are initiated to recover the missing amount.

(a) **Who are the users of the proposed system?**
(b) **List down all the possible processes, which needs to be taken up for the proposed system.**
(c) **Make a list of the external entities, which send or receive information to the system.**
(d) **Prepare a user-level context diagram that summarizes the information of MCRC.**
(e) **Draw a consolidated Context diagram for the proposed system.**

#30

The following is the process for issue of personal cheque book to customers in banks.

1. Customer fills the cheque book issue form at the counter. In the form he fills his account type, account no, name, no of cheque leaves, and signs it.
2. The clerk validates the account and checks the customer signature. If any of the information is not correct, he verifies with the customer, and if not satisfied, returns the form.
3. If details are correct, he picks the required cheque book from the stock, enters the numbers of cheque leaves against the customer account, and issues the cheque book.

Draw a structure chart to explain the sub-processes.

#31

Cinevista Ltd. is into distribution and exhibition of films produced in India. Cinevista has sub distributors appointed to handle various regions and territories within the country, through which they distribute and exhibit the movies at cinema halls. Cinevista receives a weekly report of collections from the distributors and consolidates it into an output shown below:

Collections for the week: 1/1/98 to 7/1/98

File No.	File Name	Distributor Name	Cinema Name	City/ Town	Collections in Rs.
35	Deep Impact	Shringar Films	Roxy	Bombay	5,54,788
35	Deep Impact	Shringar Films	Cine Time	Bombay	4,67,999
35	Deep Impact	Vision Track	Apollo	Delhi	3,45,125
42	Khamosh	Blaze Films	Metro	Calcutta	10,76,646
....					
....					

(a) **Normalize the above report in 3NF through 1NF and 2NF.**

(b) **Draw a functional dependency diagram for the normalized form.**

#32

CAT logistics is a central warehouse for products of Caterpillar Corporation. The corporation sends their finished goods to CAT warehouse, from where they are despatched to customers across Europe. When a sale is made for the product the corporation sent in the delivery advice to CAT containing the details of the sale. The CAT after packaging the goods delivers it to the client, and the copy of the delivery note is sent to the corporation.

The following information is part of the Despatch and Delivery note. Despatch Advice No, Despatch Advice Date, Product Code, Product Description, Order Qty, Customer Name, Customer Address, Delivery No, Delivery Date, Mode of Transport.

Draw a functional dependency diagram and obtain the normalized form.

#33

Generate a set of fully normalized tables from the following unnormalized table:

Operating schedule

Doctor Code	Doctor Name	Operation	Op. Date	Patient ID	Patient Name	Date of Adm.
54	Prasad Kamath	Bypass	4/1/99	4456	James Davis	2/1/99
13	Vikrant Goyal	Bone Marrow	4/1/99	4134	Elvis Scott	25/12/98
37	Vithal Kau	Kidney Transplant	4/1/99	4104	Evans Smith	17/12/98
....						

#34
Precision Tools employs 50 workers per shift, and works on 3 shifts of 7 Hours for 6 days. Unique number identifies each employee. This number identifies an employee card on which is written the employee's name, residential address, telephone numbers, date-of-birth and current weekly salary.

Each shift is identified by a unique shift number, which documents the start and end hours of each shift. The current shift number relevant to a given worker is updated on his card each week.

Produce a functional dependency diagram for this system. Produce a set of table structures from the dependency diagram.

#35
The following is the brief of designing an appropriate information system for the patient's appointments and operations activities of a large general hospital.

(i) Patients must make an appointment at a given clinic session held at one of the hospitals clinics.

(ii) Doctors are allocated one or more appointments within a clinic session, but only one doctor will be present at each appointment.

(iii) Operations are scheduled and allocated to one of a number of operation theater sessions held in the hospital. Each doctor may perform a number of given operations on patients. A given operation is done on only one patient, but there may be more than one doctor in attendance.

(a) List out the entities involved in this system.

(b) Establish the relationships between the entities and draw E-R diagram.

(c) Based on E-R diagram, list out the normalized data structures.

#36
Arrive at normalized data stores, based on the data in the supermarket bill illustrated in figure shown below.

A supermarket bill

Shoppers Stop		
Item	Nos.	Amount
Coffee Powder	2	76.00
Kingfisher Beer Can	5	200.00
Cosmetics	3	154.75
Soaps	6	33.00
Total 16 Items		**463.75**

Thank you for shopping at Shopper's Stop
Thane Branch - Hello 911 456786
12/3/98

#37
Arrive at normalized data stores, based on the data in the gas bill illustrated in figure shown below.

A gas bill

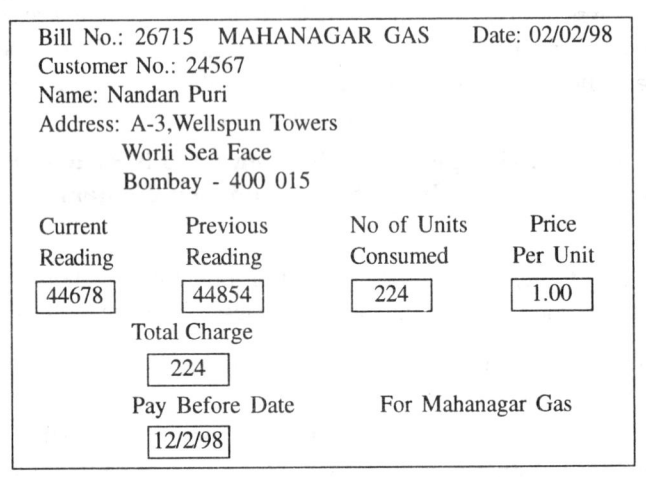

```
Bill No.: 26715   MAHANAGAR GAS      Date: 02/02/98
Customer No.: 24567
Name: Nandan Puri
Address: A-3, Wellspun Towers
         Worli Sea Face
         Bombay - 400 015

Current        Previous       No of Units    Price
Reading        Reading        Consumed       Per Unit
 44678          44854            224           1.00

         Total Charge
             224

         Pay Before Date      For Mahanagar Gas
            12/2/98
```

#38
The following is a skeleton relational schema for educational registration system.

Department (<u>dept-no</u>, dept-name, ...)
Staff Members (<u>staff-no</u>, staff-name, dept-no, ...)
Courses (<u>course-no</u>, course-name, ...)
Students (<u>student-no</u>, student-name, ...)
Allocations (<u>staff-no</u>, course-no, ...)
Registrations (<u>student-no, course-no</u>, ...)

The fields underlined indicates key fields.

Establish an Entity-Relationship model for the above structure.

#39
M/s Stable Horses are into breeding and rearing of racing horses. Following is the explanation for the function of horses breeding register, which they maintain.

A racing horse is identified by a unique name. The date of birth of the horse and its sex are also recorded. Each horse has a father and mother. The weight of the horse is monitored and recorded every month. When the horse is finally sold to the bidder, the sales details along with the purchaser details are also recorded in the register.

(a) **Produce an E-R diagram, which represents the above function.**
(b) **Produce a set of table structures from the E-R diagram.**

#40

The process of admitting people to a zoological garden every day is as follows:

A child under 3 years of age is not to be charged an admission fee. A person under 18 is to be charged half of the admission fee. If an adult accompanies a child under 12, then that adult is to be charged quarter of full admission. For persons over 18, full admission is charged, except for students who are to be charged half admission. Senior citizens (men over 65, women over 60) are to be charged quarter full admission. A discount of 10% is to apply to all persons paying full admission and are members of a party or more. There are no student concessions on the weekend.

Construct a decision table and decision tree for the above application.

#41

Salesmen in a company are paid commission for the products sold by them. The basis for paying of commissions is explained in the following para.

Shady sellers operate a graded commission policy for their salesmen. The company makes a distinction between products selling at more than Rs 5,000 and items selling under Rs 5,000. Items above Rs 5,000 are subjected to a commission of 10% if more than 300 items are sold and the salesman's salary is below Rs 20,000. Salesmen getting salaries in the region Rs 20,000–25,000 gain 8% commission, and those getting salaries above Rs 25,000, gain 7%.

If less than 300 items are sold then the commission is 8%, 7%, 6% for the same classification of salesmen. For items having a value under Rs 5,000, sales over 300 items give a flat commission of 8% to all catagories of salesmen.

Draw a decision table and decision tree for the above application.

#42

The current production controller at Brikklin describes how he produces a despatch advice as follows:

"I look through my jobs file for completed jobs—those jobs where the completion indicator is set to 'Y'. If there are enough of these to fill a trailer I'll simply despatch these. I have to do a rough calculation on the job weights, as the maximum loading on a trailer is 20 tons. If there are not enough completed jobs I'll look through the file for jobs that have partially completed; I'll then part-despatch some or all of these jobs up to the loading limit."

Produce a structured English specification of this process.

#43

Burlington Express is a mail order company specializing in supplying men's, women's and children wears fashion clothing in India. The company has tied up with some specialized textile manufactures to supply the clothing. These are based on designs provided by the Burlington, over which they have marketing rights. Each specific item is supplied by just one manufacture, but most manufactures supply many different items of clothing. Each item is allocated a unique code by the mail order company. Customers send in orders to the mail order company. Each order can contain one or more items of different categories. The mail order company orders stock from the manufactures using a clothing requisition and stores them at their warehouse. After the clothing is checked for quality and user requirements, the same is sent to despatch department for packing and despatch to customers.

(a) **Identify major entities**

(b) **List down various attributes you feel Burlington express needs to store about the entities.**

(c) **Develop an E-R diagram to model the Data thus identified.**

(d) **Obtain relations from the E-R diagram.**

#44
Establish the functional dependency for all the data elements, and generate a set of fully normalized tables from the following un-normalized table.
Property Site Visits

Property ID	Property Description	Agent Name	Potential Buyer	Date of Viewing	Expected Time for Decision (days)
P1001	Vedanta Towers	Hari Bhai	Srinivasan	4/1/97	10
			Lata	4/1/97	15
			Krishnan	6/1/97	10
			Venkat	9/1/97	30
P1030	Evershine Complex	Ajay Mehta	Paresh	4/1/97	15
			Nadkar	5/1/97	30
			Nandan	10/1/97	20
...					

#45
Local employment bureau (LEB) requires a system to hold details of the students who apply for courses run by local colleges. These courses are classified as professional, vocational, technical. Each of these courses is run by one or more colleges and has fixed number of seats. Each course and college is allocated a unique code by LEB. An applicant may apply for several courses. Details are to be kept of the standard qualifications for each course, and also the qualification's of the applicant needs to be maintained.

Based on the above scenario:
(a) **Identify the relationships between the entities.**
(b) **Draw an E-R diagram, and evolve the data structures for the system.**

#46
Dinesh Balani runs Blossom Nursery a very big plantation house which is into growing, tending and selling exotic plants all over the country. There are six such farms at different cities, where these plants are grown. Some plants, such as the more exotic varieties and trees, are bought in from different vendors around these farms. The Nursery has appointed franchise shops in metros and major cities to sell these plants. These shops also can place orders directly to their local vendors as per customer requirement for the plants not available with them.

The following are the operations at the shops.

Vendors bring the plants they have chosen to sell to the shop for payment. Currently, the shop manager works out the price of each plant in a rather arbitrary fashion. Instead the manager would like to have more information on which his prices can be based, such as the cost of materials (for example, seed and compost) and the length of time it takes for seedlings to reach the stage when they are ready for sale.

The shop also stocks a wide variety of related products, like fertilizers, seeds, compost and watering instruments. The shop manager is responsible for ordering stock, using the Nursery catalogues. All orders, including orders for the plants that bought in, are written out on a standard two-part order form. The top copy is sent to the head office and the carbon copy is placed in an order file. When goods and invoices are received the staff at the shop unpacks them, checks against the copy of the order form and prepares the cheque for payment. Normally if there is any mismatch between what has been ordered and what has been supplied, a query is sent to the supplier. The old order forms are filed in an 'orders filled' file where they are kept for two years.

The Blossom nursery wants to develop two systems, one for their farms and the other for shops. Based on the above the following are the broad requirements for the new system.

- To record data on plants, including their common and Latin names and watering instructions. (Both for farms and shops)
- To maintain accurate records of the plants grown, including the date the seed were sown, the number of trays, greenhouse in which, they are grown, the date they were estimated to be due to be ready for sale and the date they were actually removed for sale. These records are to be kept for one year. (For Farms)
- To maintain details of the suppliers of the products purchased at the shops.

- To maintain details of the suppliers of the products purchased for use in farms.
- To record data about the orders placed for products, including those for plants that are bought in, about the deliveries and invoices received and about the payments made. (Both for shops and farms)
- To aid the pricing of plants, by providing data about the length of time each type of plants takes from sowing to being ready for sale, and also about the cost of materials purchased, such as seeds and compost. (For Both shops and farms)

Based on the above, do the following in the given order.

(a) Identify all the processes, data flows and data stores. Prepare separate context diagram that summarizes the information flow for farms and shops.

(b) Convert the context diagram into DFD to explain the process and data flows to lowest level for each system.

(c) Based on the DFD output, list out the entity relationships between data stores and normalize the data structures.

(d) Convert the processes in DFD into structure charts.

(e) With Structure charts as the inputs, use process specification techniques to construct program.

#47

Tarana Music, a gramophone company established in 1950's, is into music recording and selling. The company holds information about the various records running into millions. A record consists of a number of tracks, each containing a recording of a performance of a particular artist or movie. These records are to be played at specific speeds, which are embedded, on these records and are catalogued by unique numbers based on titles. These classifications are based on *Film Music* – Indian within which regional, *Western* – Pop, Rock, Jazz, *Classical* – Indian, Western, *Folks, Devotional* and so on.

(a) Using an E-R approach to derive a data model for the gramophone record database outlined above.

#48

ACM Group is one of the top group companies, having operations on different industry sectors spread across the country. The Group Company has a subsidiary company ACM Wheels Ltd., which caters to the transport requirements of the group. The company owns a fleet of vehicles, which includes trucks, buses, cars and vans. The trucks are used to transport the goods for the various group companies, and the buses are used for employees pick up at various factory locations. Some of the cars are given to the employees for their daily use, and some are used to pick up customers/

visitors. The vans are used as pick up for local goods movement at various factories. There are four departments at ACM wheels catering to Purchase, Administration, Finance and Servicing of the vehicles. The following are the activities of each department.

The group company allocates a budget for ACM wheels for the year. The purchase department headed by Leena submits the audited expenses of the previous year on the basis of which the budget is allocated, keeping in mind the new purchases to be made, the old vehicles to be discarded and so on. The budget includes cost towards purchase of new cars, spare parts, salaries to the workforce and other administrative costs. The budget takes into account the revenue from sales of old vehicles. The company has a policy to discard all vehicles, which are more than five years old. These vehicles are either sold off to the employees through a lottery system, or bids or invited through advertisements in newspapers once a year. The purchase department on purchase of the new vehicles gets them registered and hands them over to the administration for use. The finance department settles the bill.

The administration manager headed by Krishnan looks into allocation of the vehicles for usage. These vehicles are located at the group various offices and factories. The vehicles are identified by the unique registration number and the unique employee number identifies drivers in the company. Some drivers are authorized to drive a number of vehicles. Vehicles are allocated to departments within the companies, although drivers in other department may use them. Some types of vehicles require specialist driver qualifications, which the administration department takes care at the time of recruitment. At time the existing drivers are imparted training to drive these specialized vehicles. There are occasional accidents which may lead to the vehicle being written off and/or the driver being disqualified from driving some or all classes of vehicle.

The finance department headed by Ravichandran carries out the following tasks. The department records for each vehicle identified by registration number, its current and replacement value, dates of its purchase or sale, when to claim depreciation for taxes, its warranty details, etc. For accounting purposes, the allocation of vehicles to various offices and factories is also required. Details are recorded of any insurance claims associated with accidents, repair costs if no insurance claim arose and refurnish costs. Also details are kept of regarding the dates for payment of taxes and insurance.

The service department headed by Manoj Shetty is responsible for giving regular services to each vehicle. These are different types of services corresponding to different mileage values for each class of vehicle, and the description documenting each type of service. Where vehicles have been involved in accidents, details of repairs carried out are recorded. The department also maintains the inventory of parts used frequently for servicing. If some major parts which cost above Rs. 50,000 and needs to be procured, the requirement is forwarded to the administration and finance department. Both departments jointly decide whether to purchase it or to sell off the vehicle. Based on the above descriptions:

(a) **Identify all the processes, data flows and data stores. Prepare separate context diagram that summarizes the information flow for ACM wheels.**

(b) **Convert the context diagram into DFD to explain the process and data flows to lowest level for each system.**

(c) **List out the possible entities for vehicle database covering the requirements of four departments.**

(d) **Use an E-R approach to derive a data model for the ACM wheels vehicle database from the derived data stores.**

(e) **Use Functional dependency diagram (FDD) to obtain normalized set of relations. Validate your data model obtained in (d) with FDD.**

#49

Kautilya university research group publishes an analysis of all journal papers relating to Physics, Chemistry and Mathematics. Each paper may have one or more authors, and may appear in only one journal. Journals are identified by publisher details, title, volume and issue number. One issue contains many papers on each of the objects. Each paper contains a series of reference to other papers. Authors can, and usually do, contribute to a large number of papers appearing in a variety of journals.

In order to reduce the work involved in preparing their analysis and keep track of these papers for use by scholars and research scientist, the university wishes to automate and store the relevant information on the computer. The system must record the following:

- List of all journals to which university subscribed and its publication details.
- Data of authors who regularly subscribe to these journals, along with their e-mail addresses.
- Data of authors who has written one or more papers, which discuss a particular topic.
- Where to look for the papers.

By identifying the entities, attributes and relationships for the above information, derive an appropriate data structure for the conceptual model of the research group's database.

#50

Bell U174dyog (BU) is the biggest car manufacturing company having its plant at Bilaspur (MP). The cars manufactured every day are rolled out of the factory and loaded on vehicles and sent to various dealers across the country. BU owns a fleet of vehicles, and some of them are hired as and when required by a third party fleet owners. These parties are registered with BU.

The following set of data relating to vehicles is captured from the time the cars arrive at the shipping point in the factory till they are loaded and leave the factory. The transit details are also captured. Once the cars are delivered to the dealers the BU owned vehicles return back to the plant at Bilaspur, for making further delivery.

List of complete information as captured currently:
- Vehicle no
- Capacity (no of cars/payload)
- Trip No
- First Driver name
- Second Driver name
- Arrival Bilaspur Date
- Arrival Bilaspur Time
- Arrival BU Date
- Arrival BU Time
- Departure BU Date
- Departure BU Time
- Waiting Days
- Waiting Allowance to drivers —— a
- Load Advance amount given to drivers depending upon the destination ———— b
- Diesel (in litres and amount) ———— c
- Oil (in litres and amount) ———— d
- Incentive/Reward to driver ———— e
- Gross Freight (fixed FOR wise) ———— f
- TDS by BU ———— g
- Deductions other than TDS ———— h
  Net freight ———— m = f − g + h
- Transit Repair expenses if any ———— n
- Category of transit repair
- Nature of Repair
- Bill No/Date/Vendor
- Challan in Transit ———— o
- Police expenses ———— p
- Recovery from driver ———— q
- Income ———— r

where r = m − [(a + b − q) + c + d + e + n + o + p]
- City to
- Distance in Kilometres
- Dealer name
- Unload Date
- Unload Time
- Transit Time
- Variance from BU Standard (+/-)
- Remarks – reasons for + variance
- POD Status (damaged/ok)
- Next Arrival Due date (Bilaspur)
- Return Status—empty/full
- Actual Return Arrival Due Date in Bilaspur
- Late Arrivals
- Remarks – reasons for late arrivals if any

Derive an appropriate data structure for the database relating to vehicles of Bell Udyog and its usage.

#51
A local library wants to build up a data processing system to help its libraries in the following tasks:
- Recording books lent.
- Reservation of books. It must be possible for a borrower to reserve a book in his local library, so that it is delivered as soon as it is available in any of the local authority's library.
- Printing overdue notices. The overdue notices must include the name of the borrower, the title of the book and the name of the author.
- Printing manual index card files, so that books can be searched for manually. The search argument may be the title of the book, the subject or the name of the author. Note that there may be more than one author of a single book, and that the book may deal with more than one subject.
- It must be possible to combine the search arguments on-line.

(a) Identify all the processes, data flows and data stores. Prepare separate context diagram that summarizes the information flow for library information system.

(b) Convert the context diagram into DFD to explain the process and data flows to lowest level for each system.

(c) Design an E-R diagram that describes how the library can build up the conceptual database.

(d) **What relationship types does the E-R diagram contain?**
(e) **Transform the data elements arrived so that the files of the database satisfy at least third normal form.**

#52
Explained below is the most important types of input and output for an insurance company which can be summarized as follows:
- Insurance proposals and policies include the following fields: number of the policy, annual premium and type of policy.
- Notices of claims and outgoing payments include the following: number of the policy, amount of loss and data relating to the claim.
- Premium payments include the following: number of the policy, premium and date of payment.
- The insurance agents receive index cards on clients living in their district. Each card includes an overview of the policies that a client has taken out with the insurance company.
- The letters asking for payments include the following: number of policy, overdue premium and date of payment.
- The insurance agents receive commission on new policies and on increases of premiums. The specification of commission contains the following: the name of the agent, the agent's address, commission period, and for each new policy sold in the period in question, a policy number and a commission sum. The commission sum is calculated as a percentage of the increase of the annual premium and is paid out the first time the client pays the premium of the new or increased policy. The percentage is dependent on the type of policy.
- It has to be possible to output annual statistics of the total sum of premiums per district. The system must allow for the possibility of a client changing districts.
- It has to be possible to output annual statistics of the total sum of premiums per agent. Note that an agent may change districts and that there may be more than one agent in a district.

(a) **Draw a data flow diagram to cover the above mentioned activities.**
(b) **Identify all the data elements required to create a data dictionary.**
(c) **Draw a functional dependency diagram for the identified data elements.**
(d) **Normalize the functional dependency diagram to 3NF by process of normalization.**
(e) **Validate your results by means of an E-R diagram.**

#53

Mahanagar Gas is into distribution of gas through cylinders and pipelines to household and industrial consumers in the metros. They are in a process of designing a new billing system. Meter readers go monthly to read for each residential and each business customer. The readings are used to prepare a bill that is sent to the customer. Business customers get a different kind of bill than residential customers. Payments of these bills can be settled directly through cheques at the local branch office spread across in each metro. In addition the company is also tied up with Rupee Bank, through which the customer can settle the bills. The bank everyday sends a settlement of the money collected daily from customers to Mahanagar gas.

Special meter readings are required when a customer sells a building or rents a new one, and turns over the connection possession to the new owner. These have to be calculated and billed separately to reflect a bill for part of the month. The maintenance department also provides a special reading when a meter is repaired or replaced. A separate bill is prepared in these cases.

The central accounting department of the company is provided with a daily summary of payments received and a weekly list of bills unpaid for more than 30 days. The accounting department tries to collect unpaid bills. If they are not successful, they cut off service and send the billing department a notice to terminate the account.

Based on the above, do the following in the given order.

(a) **Identify all the processes, data flows and data stores. Prepare separate context diagram that summarizes the information flow for metering of gas to consumers.**

(b) **Convert the context diagram into DFD to explain the process and data flows to lowest level for each system.**

(c) **Based on the DFD output, list out the Entity relationships between data entities and normalize the data structures.**

(d) **Convert the processes in DFD into Structure charts.**

#54

Rama opens the mail daily morning and checks orders for completeness. Incomplete orders are given to Kanhan in customer relations, who sends them back to the customer with an explanation of the problem. Complete orders are taken to Accounting department headed by Ganesha. Accounting checks the customer's credit rating. If it is ok, the order is sent to sales department, which checks to see that the item is in inventory. If it is not, sales backorders the item and informs Kanhan who informs the customer. If it is, sales prepares a three-part packing slip and sends it to shipping. Shipping pulls the items and sends them to the customer with the first copy of the packing slip. The second copy of the packing slip goes to accounting, which bills the customer. The third copy is used by sales to update the inventory records.

Based on the above, Identify all the processes, data flows and data stores. Construct a DFD that summarizes the information flow for the above mail order delivery system.

#55

Using the example of manpower search and placement agency, do the following:

(a) **Write down the various functions you feel such an agency has to carry out on regular basis. Arrange the functions in a hierarchy by means of functional decomposition diagram.**

(b) **List relevant data flows, data stores, processes and source/sinks.**

(c) **Draw a context diagram and the highest level DFD.**

(d) **Explain why you choose certain elements as processes versus sources/sinks.**

(e) **Mention the assumptions you have made while choosing data stores. Discuss various possible alternatives.**

(f) **Draw a E-R diagram to represent the data model underlying operations of the agency.**

(g) **With the help of data flows and data stores list down the important data elements to be stored in the database of the placement agency. Design a suitable scheme for naming the elements and generate a list of the variables as per the scheme.**

(h) **Mention various functional dependencies in the system and obtain a normalized database by process of normalization.**

(i) **Draw the functional dependency diagram (FDD) and use the same to obtain BCNF relations. Validate your results arrived in (h).**

(j) **List down and design various input forms required to support the operations of the placement agency.**

(k) **List down and design the key reports, you feel the agency will be generating for its management and outside world. (Assume a suitable organization structure of the agency for this purpose).**

MULTIPLE CHOICE QUESTIONS

1. The first step to the system study project is to:
   (a) Define system performance criteria
   (b) Describe information needs
   (c) Provide staff for the study project
   (d) Announce the study project.

2. During the system study, analysts determine managers' information needs by:
   (a) Conducting tours of a nearby computer centre
   (b) Asking questions
   (c) Showing samples of computer reports
   (d) Teaching short courses in programming languages.

3. Which of the following is not a factor in the failure of a systems development project?
   (a) Inadequate user involvement
   (b) Failure of systems integration
   (c) Size of company
   (d) Continuation of a project that should have been cancelled.

4. Which are the tools not used for system analysis?
   (a) System – test data
   (b) Decision table
   (c) Data flow diagram
   (d) Flowcharts.

5. The four icons used in building data flow diagram are:
   (a) Flow, Source, Store, Process
   (b) Flow, Process, Source, Store
   (c) Flow, Process, Source/Destination, Store
   (d) Source, Process, Destination, Store.

6. The first step in systems development life cycle is:
   (a) Database design
   (b) System design
   (c) Preliminary investigation and analysis
   (d) Graphical user interface.

7. Which of the following is not considered as a tool at the system design phase?

(a) Pie chart
(b) Data flow diagram
(c) Decision table
(d) Systems flowchart.

8. A pseudo code is:
   (a) A machine code
   (b) A computer generated random number
   (c) A protocol used in data communication
   (d) Easy way to communicate the logic of a program, in English-like statements.

9. Data dictionary contains detail of:
   (a) Data structures
   (b) Data flows
   (c) Data stores
   (d) All of the above.

10. Decision tree uses:
    (a) Pictorial depiction of alternate conditions
    (b) Nodes and branches
    (c) Consequences of various depicted alternates
    (d) All of the above.

11. Which of the following tools is not used in modeling the new system?
    (a) Decision tables
    (b) Data dictionary
    (c) Data-flow diagrams
    (d) Process descriptions.

12. Which of the following tools are used for process description except:
    (a) Structured English
    (b) Decision tables
    (c) Decision trees
    (d) Data dictionaries.

13. A feasibility document should contain all of the following except:
    (a) Project name
    (b) Problem descriptions

(c) Feasible alternative
(d) Data-flow diagrams.

14. In DFD, an originator or receiver of the data is usually designated by
    (a) A circle
    (b) An arrow
    (c) A square box
    (d) A rectangle.

15. Difference between Decision-Tables and Decision Trees is(are):
    (a) Value to end user
    (b) Form of representation
    (c) One shows the logic while other shows the process
    (d) All of the above.

16. The main purpose of the system investigation phase is to produce:
    (a) A design report
    (b) A requirement report
    (c) A feasibility report
    (d) All of the above.

17. A decision table facilities conditions to be related to:
    (a) Actions
    (b) Programs
    (c) Tables
    (d) Operation.

18. A structure chart is:
    (a) A document of what has to be accomplished
    (b) A hierarchical partitioning of the program
    (c) A statement of information processing requirements
    (d) All of the above.

19. A decision table, part of process specification:
    (a) Represents the information flow
    (b) Documents rules, that select one or more actions, based on one or more conditions from a set of possible conditions
    (c) Gets an accurate picture of the system
    (d) Shows decision paths.

20. A data flow diagram is:
    (a) The modern version of flowchart

(b) Mainly used at the systems specification stage
(c) The primary output of the systems design phase
(d) All of the above.

21. During what phase, the requirement analysis is performed?
    (a) System design phase
    (b) System development phase
    (c) System analysis phase
    (d) System investigation phase.

22. In decision trees:
    (a) Nodes represent the conditions, with the right side of tree listing the actions to be taken
    (b) Root is drawn on the left and is the starting point on the decision sequence
    (c) The branch depends on the condition and decisions to be made
    (d) All of the above.

23. The data flow diagram shows:
    (a) The flow of data
    (b) The processes
    (c) The areas where they are stored
    (d) All of the above.

24. Which of the following symbols is (are) not the data flow diagrams (DFD)?
    (a) A square
    (b) An open rectangle
    (c) A circle
    (d) A triangle
    (e) A bubble.

25. The rule(s) to follow in constructing decision tables is (are):
    (a) A decision should be given a name
    (b) The logic of the table is independent of the sequence in which condition rules are written, but the action takes place in the order in which the events occur
    (c) Standardized language must be used consistently
    (d) All of the above.

26. All of the following tools are used for process description except:
    (a) Data dictionaries

(b) Structured English
(c) Decision tables
(d) Pseudo code
(e) All of the above are used.

27. Which of the following is not true of the conversion phase of the development life cycle?
    (a) The user and systems personnel must work closely together
    (b) Steps must be taken to phase out the old system
    (c) Documentation should be emphasized
    (d) The non-machine components of the system should be considered
    (e) All of the above are true.

28. Which of the following is done in order to gather data in phase 1 of the SDLC?
    (a) Conducting interviews
    (b) Observing operations
    (c) Using questionnaires to conduct surveys
    (d) Reviewing policies and procedures
    (e) All of the above.

29. In phase 1 of the SDLC, which of the following aspects are usually analyzed?
    (a) Outputs
    (b) Inputs (transactions)
    (c) Controls
    (d) Existing hardware and software
    (e) All of the above.

30. Which of the following is not a phase in the SDLC?
    (a) Analyze current system
    (b) Define the latest technology
    (c) Design new system
    (d) Develop and implement new system
    (e) None of the above.

TRUE/FALSE QUESTIONS

1. Structured English is easier to convert to program code than regular narrative English.

2. Data store reflects a data structure at rest whereas data flow is a data structure in motion.
3. A systems flow chart is integral to program documentation.
4. Only the design phase of the system life cycle produces documentation.
5. Data flow diagrams are drawn only as a part of a systems design document and have no relevance at the stage of systems analysis.
6. Data flow diagrams are useful in representing the decision process in the system.
7. User interaction is required only during the system study and analysis phases.
8. Data dictionaries can be used for detecting errors.
9. Data dictionary in conjunction with DFD provides an important part of the local bridge between analysis and design.
10. Whereas system design defines what is to be done, system analysis tells how it is to be done.
11. Structured flowcharts and structure charts are one and the same.
12. Data flow diagrams contain decision trees.
13. Decision trees and decision tables perform the same function.
14. A system flowchart is not a part of a program documentation package.
15. The system review phase is carried out periodically even after the system has been successfully implemented.
16. For most people decision trees are easier to understand than decision tables.
17. Processing logic can be shown only by decision tables.
18. An airline reservation system is a batch processing system.
19. The real time system is a particular case of an on-line system.
20. Data dictionary is a structured repository of data about data.
21. In decision tables, for each rule there can be more than one condition being true.
22. A data flow diagram is the same thing as a flow chart.
23. A structured chart is a sequential representation of program design.
24. In a Logical DFD the flow are restricted to show the movement of data only.
25. The Decision Tree is a pictorial representation of the sequence of program execution.
26. Entity Relationship Diagrams are used to design files.
27. Systems Overview can be prescribed by data flow diagram.
28. A data dictionary is a listing of all data elements in a data base.

29. In data flow diagrams, an arrow portrays a data flow.
30. A data flow diagram is the primary tool used in structured system development to graphically depict systems.
31. In data flow diagram an originator or receiver of data is usually designated by a rectangle.
32. Decision tables and decision trees are alternative tools for defining data process, but decision trees are easier for most people to understand.
33. A structure chart is a hierarchical diagram showing the relationship between various program modules.
34. Decision table is considered as a tool at the system design phase.
35. A data dictionary contains information about and definitions of data used in a system.
36. Structured English is often called pseudo-code because of its similarity to program code.
37. A structure chart is a statement of information processing requirement.
38. A data flow diagram is the primary output of the systems design phase.
39. A triangle symbol is used in the data flow diagram.
40. A square symbols defines a source or destination of system data.

Appendix-A

STRUCTURED METHODOLOGY ELEMENTS

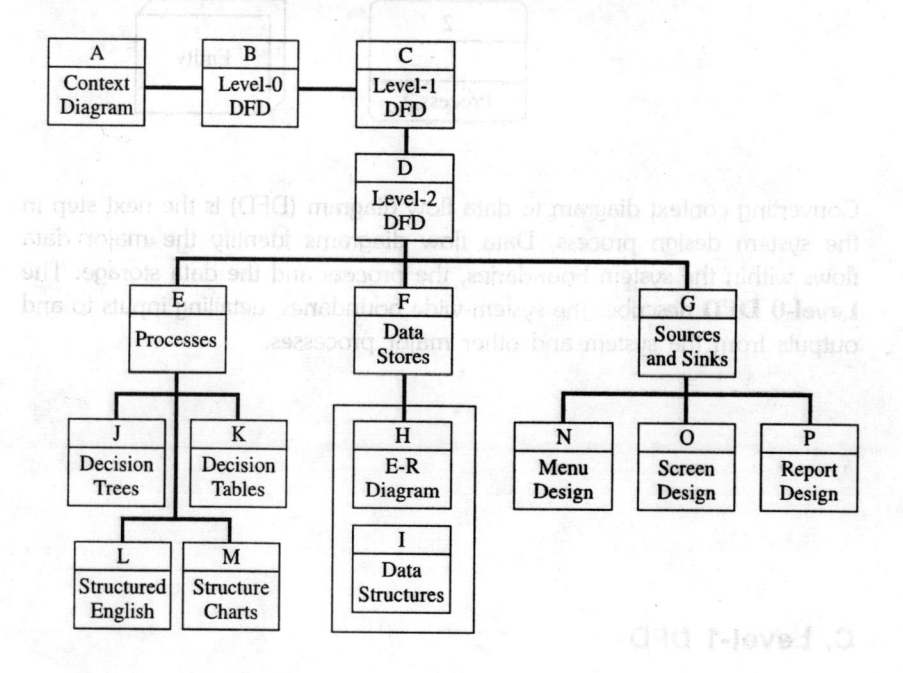

```
    A              B              C
Context        Level-0        Level-1
Diagram         DFD            DFD
                                │
                                D
                             Level-2
                              DFD
    ┌──────────────┬──────────────┐
    E              F              G
Processes      Data           Sources
               Stores         and Sinks
    │              │              │
 ┌──┴──┐           H        ┌─────┼─────┐
 J     K         E-R        N     O     P
Decision Decision Diagram  Menu Screen Report
Trees   Tables             Design Design Design
 │                I
┌┴┐            Data
L  M           Structures
Structured Structure
English   Charts
```

A. Context Diagram

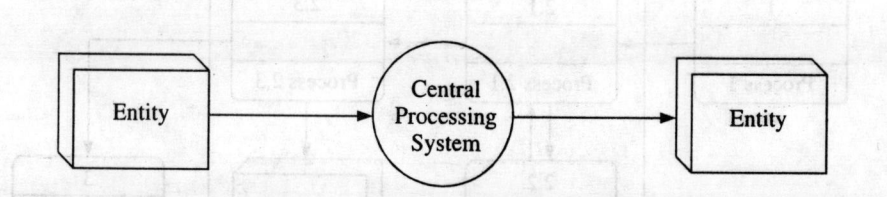

Entity → Central Processing System → Entity

The first of the activities in analyst initiates is working on the **context diagram**. The context diagram is a structured tool that focuses more specifically on system requirements and boundaries. The real power here is that the scope of the entire system can be understood at a glance; also input and output which are going to become interface are identified.

B. Level-0 DFD

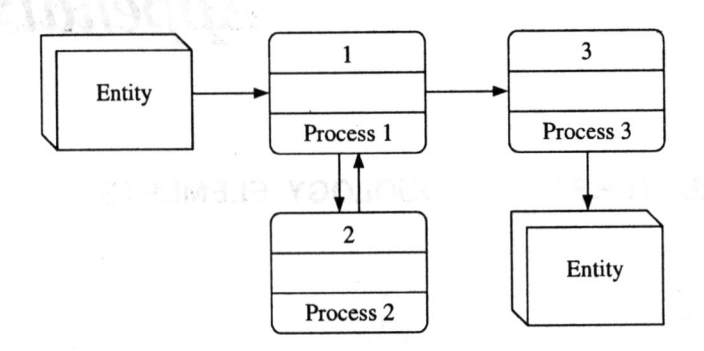

Converting context diagram to data flow diagram (DFD) is the next step in the system design process. Data flow diagrams identify the major data flows within the system boundaries, the process and the data storage. The **Level-0 DFD** describes the system-wide boundaries, detailing inputs to and outputs from the system and other major processes.

C. Level-1 DFD

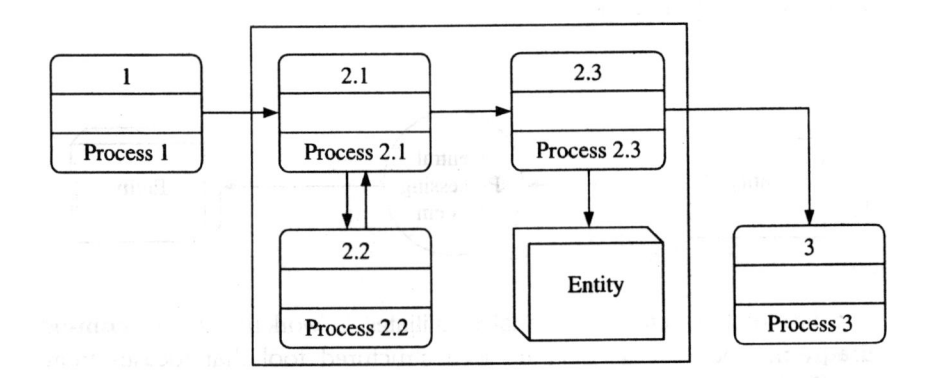

A **Level-1 DFD** describes the next level of detail within the system, elaborating flows between subsystems, which make up the whole system under consideration.

D. Level-2 DFD

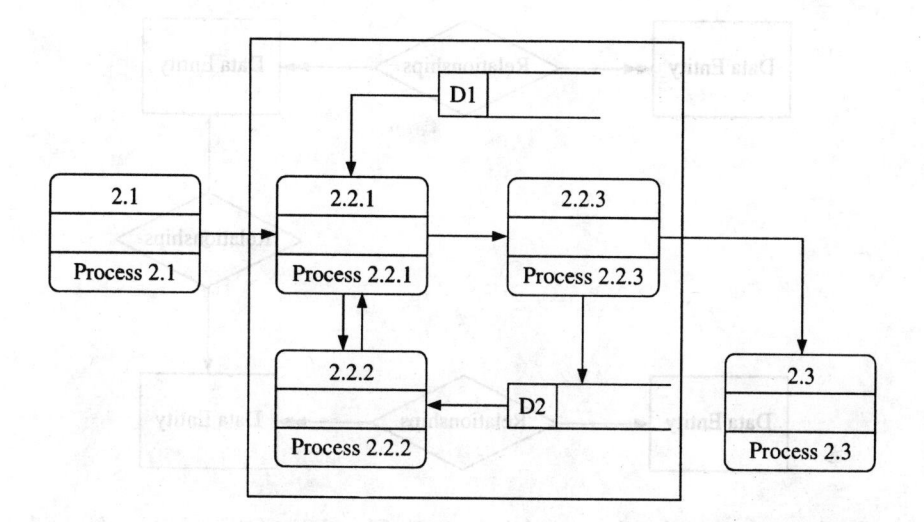

```
         ┌─────────────────────────────────────────┐
         │              ┌──────┐                   │
         │              │  D1  │                   │
         │              └──────┘                   │
         │  ┌─────┐   ┌───────┐   ┌───────┐        │
         │  │ 2.1 │──▶│ 2.2.1 │──▶│ 2.2.3 │        │
         │  ├─────┤   ├───────┤   ├───────┤        │
         │  │Proc │   │Process│   │Process│────────┼──┐
         │  │ 2.1 │   │ 2.2.1 │   │ 2.2.3 │        │  │
         │  └─────┘   └───────┘   └───────┘        │  │
         │              │  ▲                       │  │
         │              ▼  │                       │  ▼
         │           ┌───────┐  ┌──┐           ┌───────┐
         │           │ 2.2.2 │◀─│D2│           │  2.3  │
         │           ├───────┤  └──┘           ├───────┤
         │           │Process│                 │Process│
         │           │ 2.2.2 │                 │  2.3  │
         │           └───────┘                 └───────┘
         └─────────────────────────────────────────┘
```

Finally, the **Level-2 DFD** details the files in which the data is stored in the system, and from which data is obtained. In Level-2 each individual process is shown to elementary level.

E. Processes   F. Data Stores

```
┌──────────────────┐      ┌──────────────────┐
│ All elementary   │      │ All elementary   │
│ level DFD        │      │ level DFD        │
│ Processes        │      │ Data Stores      │
│   ┌─────────┐    │      │                  │
│   │  2.2.2  │    │      │   ┌────┬─────┐   │
│   ├─────────┤    │      │   │    │     │   │
│   │Process  │    │      │   └────┴─────┘   │
│   │ 2.2.2   │    │      │                  │
│   └─────────┘    │      │                  │
└──────────────────┘      └──────────────────┘
```

G. Sources and Sinks

```
        ┌──────────────────┐
        │ All elementary   │
        │ level            │
        │ Sources and Sinks│
        │                  │
        │      ▱▱▱         │
        │                  │
        └──────────────────┘
```

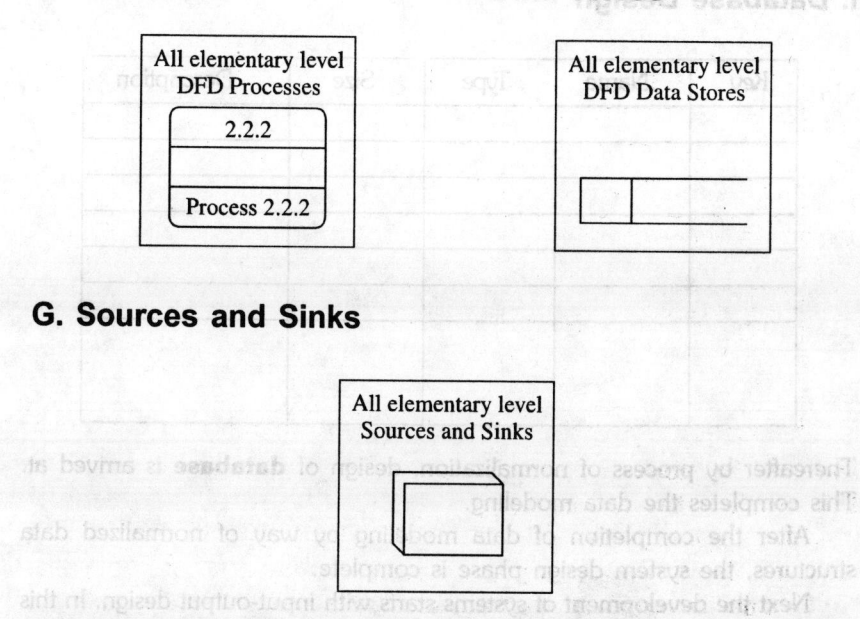

The DFD activity helps in establishing the **processes, data stores, sources and sinks,** and completes the application modeling.

H. E-R Diagram

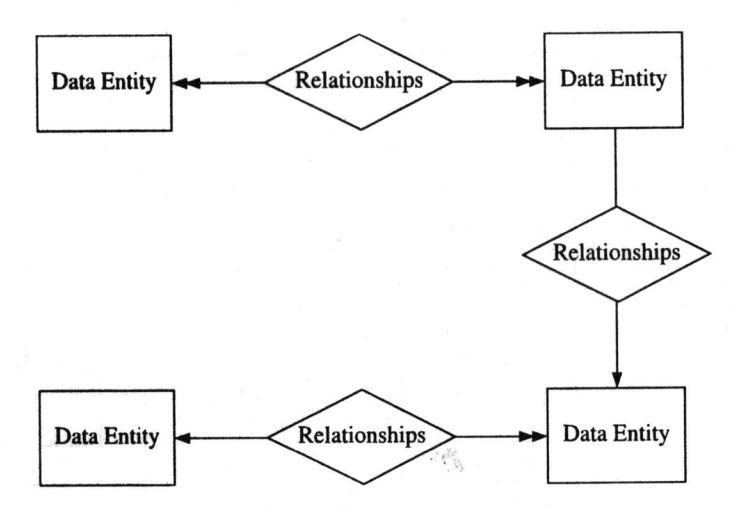

```
[Data Entity] ◄──< Relationships >──► [Data Entity]
                                            │
                                      < Relationships >
                                            │
                                            ▼
[Data Entity] ◄──< Relationships >──► [Data Entity]
```

In the next step the data modeling starts. The data stores are transformed into **E-R diagram** where the relationships that exist between the data entities are established.

I. Database Design

Key	Name	Type	Size	Description

Thereafter by process of normalization, design of **database** is arrived at. This completes the data modeling.

After the completion of data modeling by way of normalized data structures, the system design phase is complete.

Next the development of systems starts with input-output design. In this phase the analyst based on sources and sinks, the **menu, screens and reports design** for the process are carried out.

J. Decision Trees

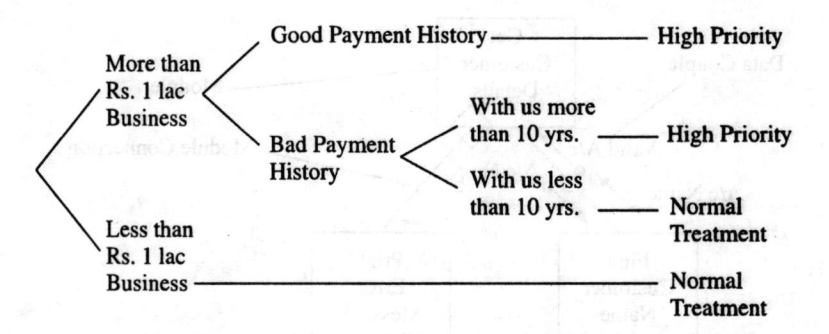

```
                    Good Payment History ─────────── High Priority
        More than  /
        Rs. 1 lac /
        Business  \                    With us more
                   \                  / than 10 yrs. ─── High Priority
                    Bad Payment  ────<
                    History          \ With us less
                                       than 10 yrs. ─── Normal
       /                                                Treatment
      /
      \ Less than
        Rs. 1 lac
        Business ──────────────────────────────────── Normal
                                                      Treatment
```

K. Decision Tables

Current balance >= 1000	Y	Y	Y	Y	N	N	N	N
Number of overdrafts <= 2	Y	Y	N	N	Y	Y	N	N
Average savings balance >= 500	Y	N	Y	N	Y	N	Y	N
Approve	X	X						
Cond. approve			X		X			
Reject				X		X	X	X

L. Structured English

Example
IF customer does more than X business
   And IF (customer has good payment history)
      THEN priority treatment
   ELSE (bad payment history)
      IF customer doing business for more than Y years
         THEN priority treatment
      ELSE
         Normal treatment
ELSE
   Normal treatment

M. Structure Charts

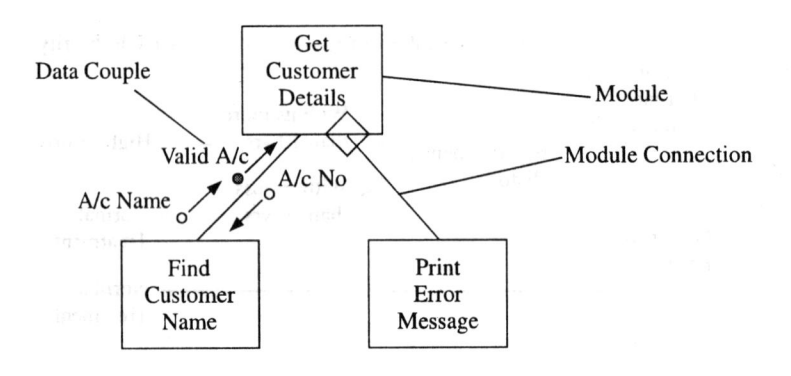

```
                    Get
Data Couple      Customer
                  Details                      ── Module
              Valid A/c
                      A/c No                ── Module Connection
   A/c Name
           
        Find                Print
      Customer              Error
        Name                Message
```

N. Menu Design

ORDER TRACKING MAIN MENU

Options:

1. Enter Orders
2. Enter Invoices
3. Print Order Register Report
4. Exit to Main Menu
5. Help for Order Tracking Process

ENTER YOUR OPTION: 5

The menus, reports and screens of a system are the user interface. The user interface is the only part of the system that the user sees. Hence, the user interface is the most important part of the system to the user. Its design must be such that in addition of providing information to the user it should look appealing.

Finally, for all the processes identified in the system, the process specifications start. This is done by way of **structure charts, decision trees, decision tables and structured English.**

Process specifications supplement the structured and graphic techniques of DFD to specify the process logic. While process specifications are normally used for lowest-level processes, it is effectively used to prepare process descriptions that cannot be shown on data flow diagrams.

O. Screen Design

```
Run Date xx/xx/xx          ORDER ENTRY SCREEN                Screen:1/2


ORDER NUMBER      [        ]      ORDER DATE (mm/dd/yy)   [  ][  ][  ]
CUSTOMER NUMBER   [        ]      CUSTOMER NAME
ORDER STATUS CODE [   ]           [                              ]
ORDER ITEMS:
        ITEM NO      PRODUCT NO       QTY ORDERED      UNIT PRICE
           1         [        ]       [        ]       [        ]
           2         [        ]       [        ]       [        ]
           3         [        ]       [        ]       [        ]

   Press F1 for Help of Customer Code
   Press F2 for Product Number
   Press F5 for Next Screen
   Press F10 to exit

     MESSAGE:   [Product No not defined in Product Master - Enter to continue]
```

P. Report Design

AAA	BBB	DDD	EEE
ZZ9	XXX	XXXXXXXXXXXXXXXXXXXXXXX	Rs. Z,ZZ,ZZZ.99
ZZ9	XXX	XXXXXXXXXXXXXXXXXXXXXXX	Rs. Z,ZZ,ZZZ.99
ZZ9	XXX	XXXXXXXXXXXXXXXXXXXXXXX	Rs. Z,ZZ,ZZZ.99
ZZ9	XXX	XXXXXXXXXXXXXXXXXXXXXXX	Rs. Z,ZZ,ZZZ.99
			Rs. Z,ZZ,ZZ,ZZZ.99

Appendix-B

WEB CASE STUDY

Overview

Pentagon Systems, one of the major Internet service provider, has decided to develop and launch portals as part of their diversification process. In the first of such portals, Pentagon has identified portal dedicated to online auctioning. The portal *myShop.com* will provide a fast and convenient way to connect buyers and sellers through Internet based on-line auctioning forum.

The portal is designed to benefit for both bidders and sellers. The portal must provide a channel for connecting buyers and sellers through an on-line auctioning forum. The services provided by *myShop.com* portal enable sellers to show case their products, on which the users will place their bid through the portal. The seller will post details of item and specifies the reserved price for each item. Pentagon will charge the seller based on the dimension of the web page. The page will be on-line for the period agreed upon by the seller, after which the seller, if required, can renew it. In the beginning the portal will only facilitate in accepting bids and pass it on to the seller. Once the seller decides and accepts the bid, the results are displayed back on the site, with necessary details of bids placed. Pentagon Systems is not involved in the actual transactions between buyers and sellers.

As the first step towards systems design, an Organization-Level Context diagram (in this case portal specific) is drawn out to identify the general requirements for the portal myShop.com as shown in Figure B-1.

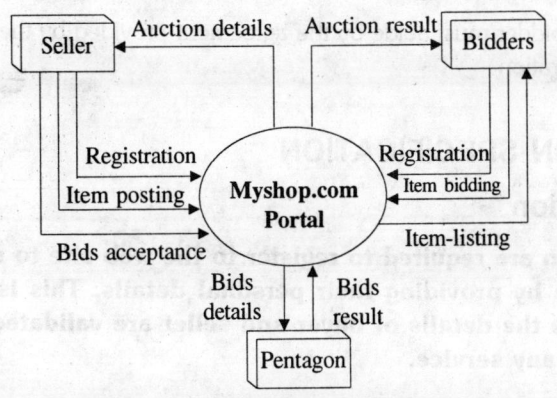

Figure B-1 Context diagram for myShop.com.

Based on the context diagram the requirement specification was worked out for the *myShop.com* portal, which is listed down. The processes to be covered are:

- **Registration**
  User registrations for using the web site.
- **Item posting**
  Registered seller to post item for sales.
- **Bidding**
  Registered user to buy the item, which is available for sales.
- **Item Listing**
  Listing of all the items which is available for sale, as per the category.
- **Auction Management**
  Send e-mail, to the bid winner and to the seller.

Initially, Pentagon system does not want the following functions to be covered in the initially development phase.

- Payment process
- Item shipments
- Payments settlement can be made on-line
- Advertising Facility can be provided on-site
- No verification of registered users
- Pentagon will have no ownership of the information provided by buyer or seller
- Item registered for auction are assumed to be as shown in the portal page
- Customer who is bidding for an item is really interested and capable of buying the same.
- The bidder must abide by the agreement provided by the portal during registration

FUNCTION SPECIFICATION

Registration

The Netizen are required to register to the web site to make use of the service by providing their personal details. This is mandatory function as the details of buyer and seller are validated before the site offers any service.

Process

From the home page of the portal, user invokes the Registration button. In the registration form the user details are captured and after the user submits, the data is routed to the web server where it is validated and updated in the database. On successful addition user is send the confirmation on-line.

Input

Personal details of the user, User name, User e-mail, Address, Phone no, City, Country, Password etc.

Output

Confirmation of registration (Confirmation or fail message page).

Item Posting

The registered user of site can make their item available for auctioning by posting details of items.

Process

After validation by the site, the user receives an item posting form in which he fills details of items he wants to auction and submits it back to system. The web server is supposed to post item details in database after validation and send back a confirmation message to user. If incomplete form is filled, the user is asked to post the items again.

Input

Item details – Item name, description, start date and last date of receiving bids, reserved price.

Output

Confirmation of item posting (Confirmation or fail message page).

Bidding

The registered user of *myShop.com* can bid for any item available on site for auctioning. The user can view the category wise item details and quote his price for item.

Process

After validation by the site, the user receives a bidding form in which he fills his price for that item and submits it back. The web server after validation will update the bid details in database, and sends back a confirmation message to user.

Input
User bid price for Item

Output
Confirmation of bid posting (Confirmation or fail message page).

Auction Result
This function is used to send the information to both seller and buyer (higher bidder) for particular item once the last date of bidding is over.

Process
It checks database for items for which the last date of auctioning is over, and sends e-mail to the seller giving details of the bids received, highest bid price and email of the bidder. If there is no bidder for item or the bid price is less than the reserved price, the seller is informed accordingly and the database is updated.

Input
NA

Output
E-mail to seller and buyer as mentioned above.

Item Search
This function is used to create dynamic list of items available for auctioning on various criteria like category, sellers and buyers location, item price range, etc.

Process
It checks database for items details according to user requirement and create dynamic list of items as per fulfilment and send back it to user so that he can bid for a particular item of his interest.

Input
Search parameter for items (like name, description, price range, category and location)

Output
Item list as per required criteria.

Based on the above major processes, next we draw the data flow diagram. Figures B-2–B-5 show the Level-1 DFD.

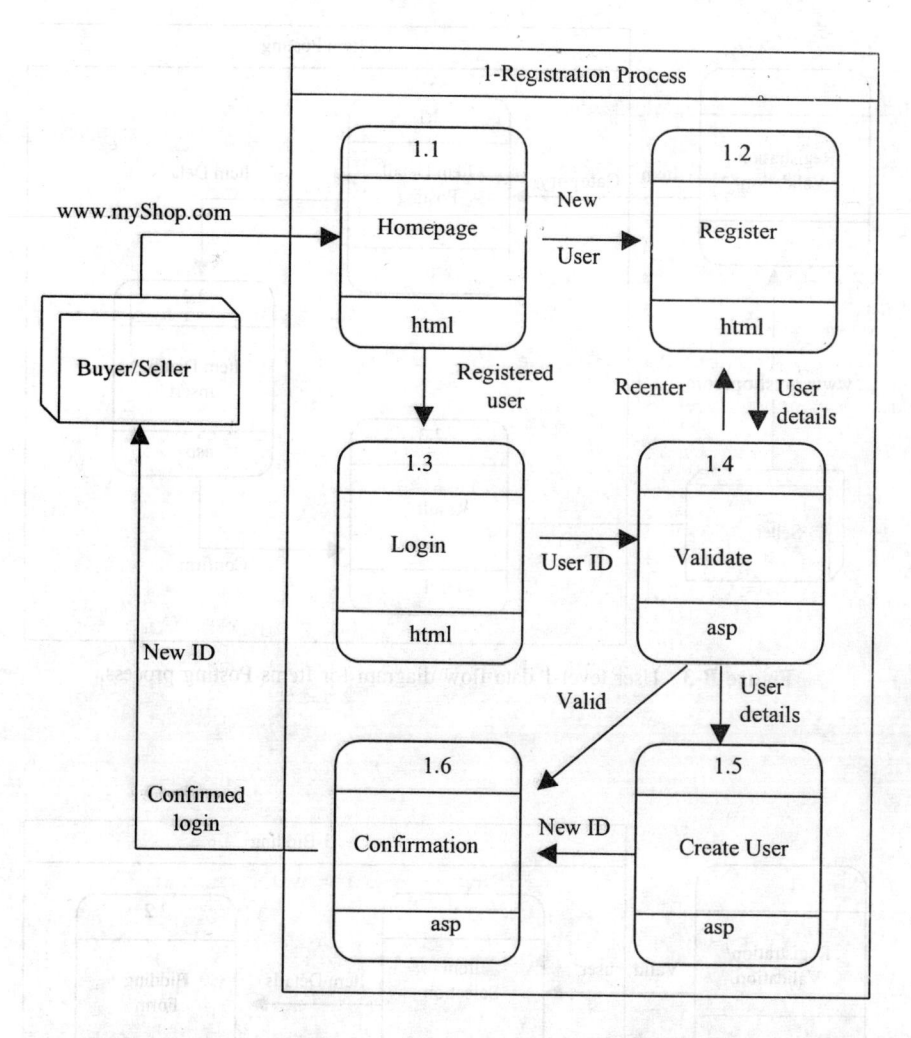

```
                    ┌─────────────────────────────────────────────┐
                    │           1-Registration Process            │
                    │                                             │
                    │   ┌─────────┐              ┌─────────┐      │
                    │   │  1.1    │    New       │  1.2    │      │
 www.myShop.com     │   │Homepage │────User─────▶│Register │      │
──────────────────▶ │   │         │              │         │      │
                    │   │  html   │              │  html   │      │
  ┌──────────┐      │   └────┬────┘              └─────────┘      │
  │Buyer/    │      │        │ Registered    Reenter ▲  │ User    │
  │Seller    │      │        │ user                  │  ▼ details │
  └──────────┘      │   ┌────▼────┐              ┌─────────┐      │
        ▲           │   │  1.3    │              │  1.4    │      │
        │           │   │         │   User ID    │         │      │
        │ New ID    │   │  Login  │─────────────▶│Validate │      │
        │           │   │         │              │         │      │
        │           │   │  html   │       Valid  │  asp    │      │
        │           │   └─────────┘         ┌────┴─────────┘      │
        │           │                       │         │ User      │
        │           │   ┌─────────┐         │         ▼ details   │
        │ Confirmed │   │  1.6    │         │    ┌─────────┐      │
        │ login     │   │         │ New ID  │    │  1.5    │      │
        └───────────┼───│Confirm- │◀────────┘    │Create   │      │
                    │   │ation    │              │ User    │      │
                    │   │  asp    │              │  asp    │      │
                    │   └─────────┘              └─────────┘      │
                    └─────────────────────────────────────────────┘
```

Figure B-2 User level-1 data flow diagram for Registration process.

DESIGN SPECIFICATION

In web applications, there are basically two types of pages: HTML and ASP. The information viewed by the user does not changes with each log in, they are coded as HTML page, whereas when the information is to be provided dynamically during access, it is coded in ASP.

In Figures B-2, B-3, B-4 and B-5 each of the processes are identified if it is an HTML or an ASP and accordingly the process specifications are worked out.

222 *Workbook on Systems Analysis and Design*

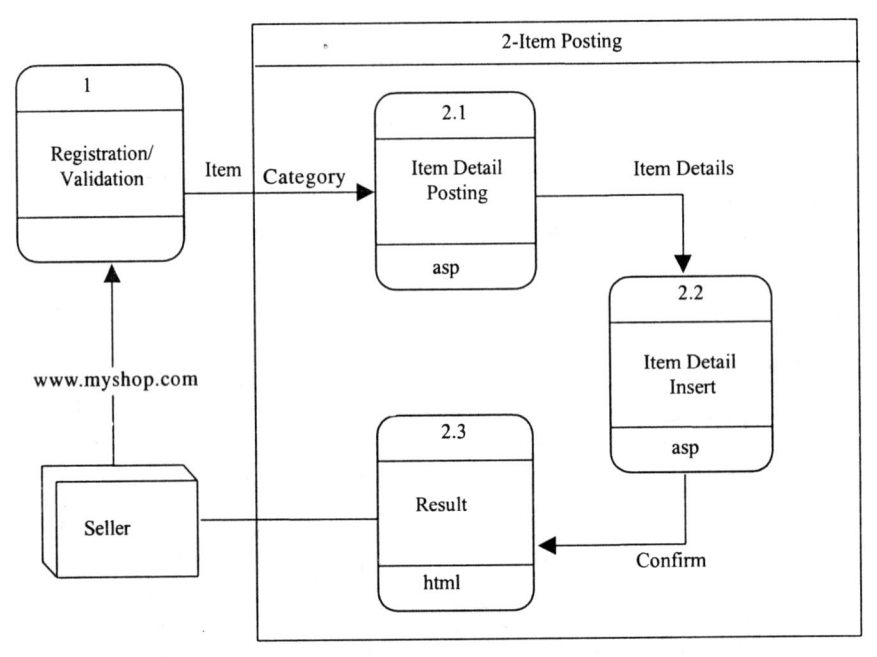

Figure B-3 User level-1 data flow diagram for Items Posting process.

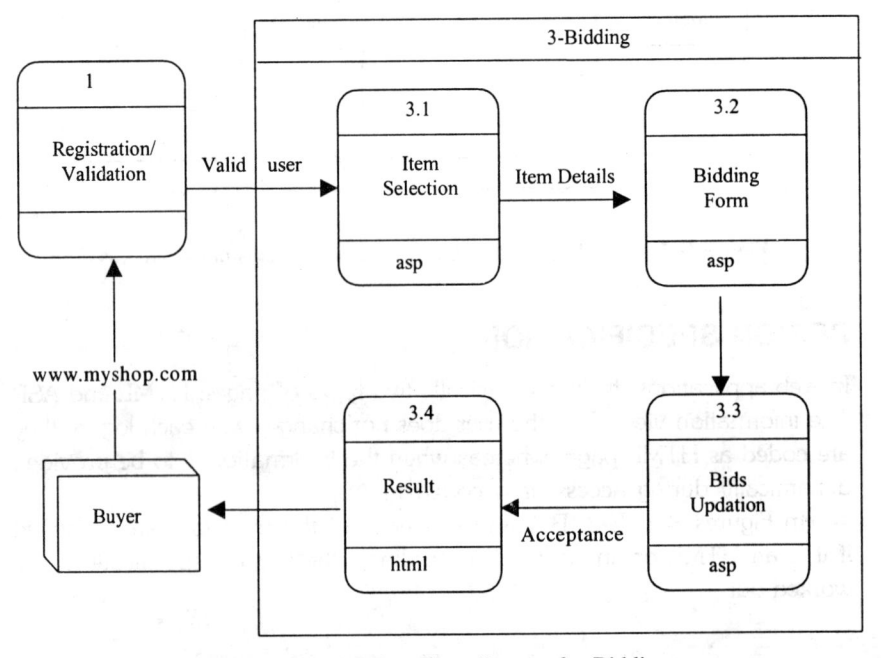

Figure B-4 User level-1 data flow diagram for Bidding process.

Appendix-B **223**

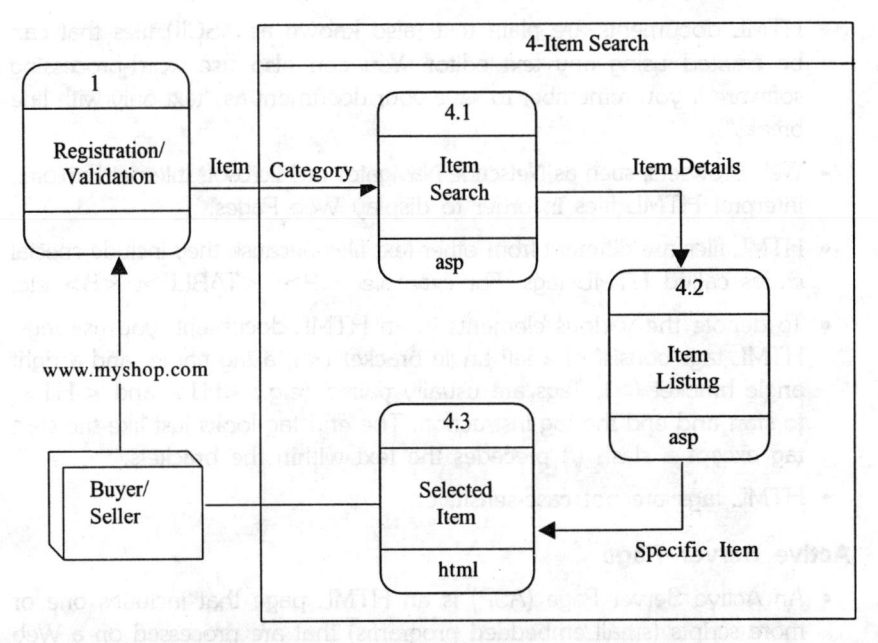

Figure B-5 User level-1 data flow diagram for Item Search.

HTML vs ASP

HTML	ASP
HTML is the *lingua franca* for publishing hypertext on the World Wide Web. It is a non-proprietary format based upon SGML, and can be created and processed by a wide range of tools, from simple plain text editors—you type it in from scratch—to sophisticated WYSIWYG authoring tools	An Active Server Page (ASP) is an HTML page that includes one or more scripts (small embedded programs) that are processed on a Microsoft Web server before the page is sent to the user. An ASP is somewhat similar to a server-side include or a common gateway interface (CGI) application in that all involve programs that run on the server, usually tailoring a page for the user
HTML Page run on the Browser	ASP Page run on the Web Server
It is a Static Page	It is a Dynamic Page

Hyper Text Markup Language

- HTML is the *lingua franca* for publishing hypertext on the World Wide Web. It is a non-proprietary format based upon SGML, and can be created and processed by a wide range of tools, from simple plain text editors—you type it in from scratch—to sophisticated WYSIWYG authoring tools.

- HTML documents are plain text (also known as ASCII) files that can be created using any text editor. You can also use word-processing software if you remember to save your document as "text only with line breaks"
- Web Browsers, such as Netscape Navigator or Microsoft Internet Explorer, interpret HTML files in order to display Web Pages.
- HTML files are different from other text files because they include special codes called HTML tags. For example, <P>, <TABLE>, <B> etc.
- To denote the various elements in an HTML document, you use *tags*. HTML tags consist of a left angle bracket (<), a tag name, and a right angle bracket (>). Tags are usually paired (e.g., <H1> and </H1>) to start and end the tag instruction. The end tag looks just like the start tag except a slash (/) precedes the text within the brackets.
- HTML tags are not case-sensitive.

Active Server Page
- An Active Server Page (ASP) is an HTML page that includes one or more scripts (small embedded programs) that are processed on a Web server before the page is sent to the user.
- An ASP is somewhat similar to a server-side include or a common gateway interface (CGI) application in that all involve programs that run on the server, usually tailoring a page for the user.
- An ASP can contain server-side scripts. By including server-side scripts, you can create Web Pages with dynamic content.
- An ASP provides a number in build-in objects. By using the built-in objects accessible in an ASP, you can make your scripts much more powerful. These objects allow you to both retrieve information from and send information to browsers.
- ASP comes bundled with a number of standard server-side Active X components like Ad Rotator, Page counter etc.
- An ASP can interact with a Database such as Microsoft SQL Server, Oracle etc.
- Using ASP we can create personalized Web Pages that display different content to different users.

Based on the above differences between HTML and ASP, we can identify the HTML and ASP pages, and describe each of the processes.

HomePage.Html

This is an HTML page, which is the front page of the portal, and guides the user to navigate the portal.
The forms submit various actions depending upon the selection.

SellerLogin.Html

This is an HTML page, which contains a form for logging in of user-id and password. When the user fills up the form and submits it, the page checks for client side validation before submitting it to web server.
The forms submit action is SellerValidation.asp page and submit method used is Post method.

BuyerLogin.Html

This is an HTML page which contains a form for logging in of user-id and password. When the user fills up the form and submits it, the page checks for client side validation before submitting it to web server.
The forms submit action is BuyerValidation.asp page and submit method used is Post method.

Register.Html

This is an HTML page, which asks the user to initiate the registration process.
The page submits action is RegValidation.asp.

RegValidation.Asp

This is an ASP, which user has entered on the register.html page. This ASP checks for existence of user in user table. If the user is not registered it sends back the user on register.html page with reasons.
Once the user is registered it invokes Usercreate.asp.

UserCreate.Asp

This is an ASP, which matches user password from Register.Html page and RegValidation.asp page.
If the user details match, it updates the user table and redirects it to UserConfirm.asp page, otherwise it send a message of mismatch and redirects to RegValidation.asp page.

UserConfirm.Asp

This is an ASP, which dynamically creates welcome message for user registration successfully by confirming name and user id of the user.
This contains a hyperlink to HomePage.asp.

SellerValidation.Asp

This is an ASP, which validates user in user table for user identification and password.

'If valid' stores user identification in a session variable and redirects ItemPostDetail.asp, else it forces the user to Register.html page.

ItemPostDetails.Asp

This is an HTML page, which contains a form to input item details which the user wants to post for auction.
The form submits action is ItemInsert.asp.

ItemInsert.Asp

This is an ASP which inserts auctioneer posted item details into Item table. *If inserted successfully it redirects to ItemConfirm.asp, else sends a message of posting failure and redirects to ItemPostDetails.asp page.*

ItemConfirm.Asp

This is an ASP which dynamically creates welcome message for successful item posting.
Contain a hyperlink to HomePage.asp page.

BuyerValidation.Asp

This is an ASP which validates user in user table for user id and password. *'If valid' then it store user id in a session variable and redirects to Listing.Asp page otherwise it forces the user to Register.Html page*

ItemBid.Asp

This is an ASP which access item details of user selected item from item table and generates a form for bidding. The user enters his bid price and posts it.
The form submits action is BidPost.asp.

BidPost.Asp

This is an ASP which updates item bid price and bidder details Item table. *'if successfully updated' redirects to BidConfirm.Asp else sends message of item bidding failure and redirect to ItemBid.Asp page.*

BidConfirm.Asp

This is an ASP which dynamically creates welcome message for item bidding process to be successful.
Contains hyperlink to HomePage.asp.

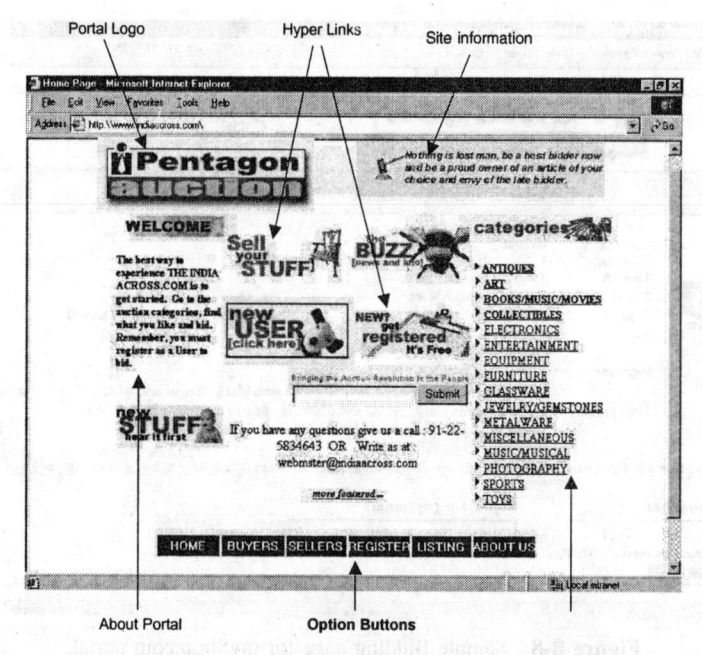

Figure B-6 Sample Home page for myShop.com portal.

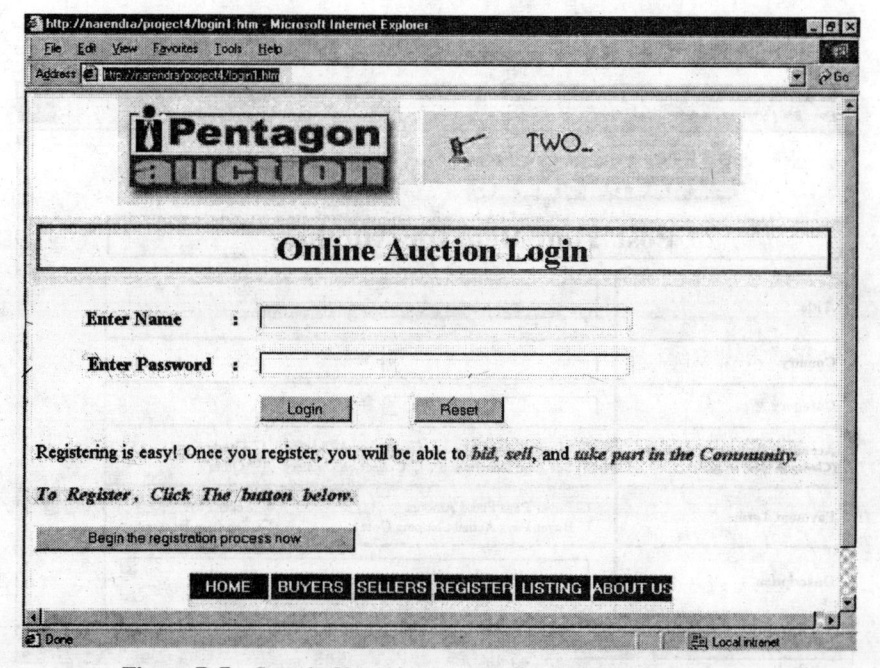

Figure B-7 Sample User login screen for myShop.com portal.

228 *Workbook on Systems Analysis and Design*

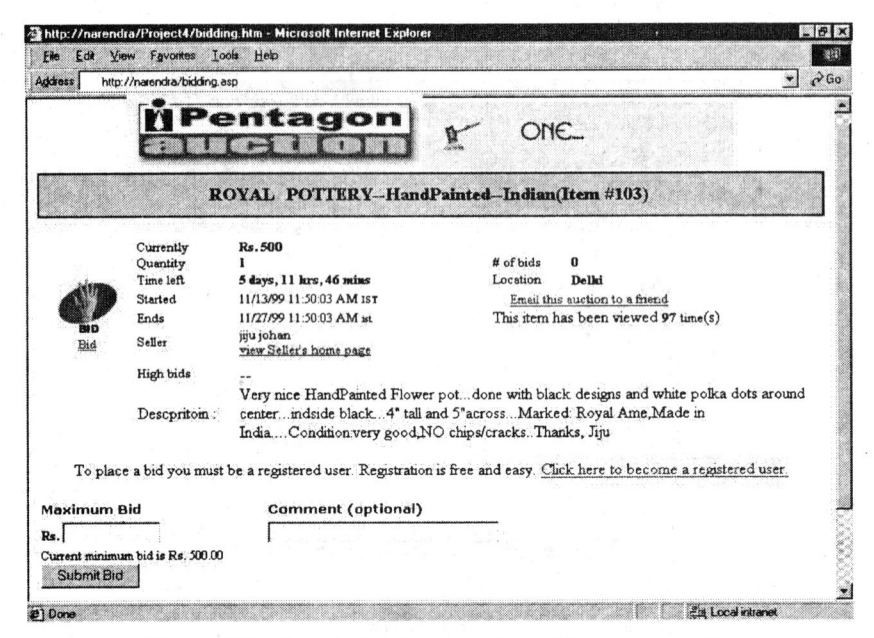

Figure B-8 Sample Bidding page for myShop.com portal.

Figure B-9 Sample Auction item posting page for myShop.com portal.

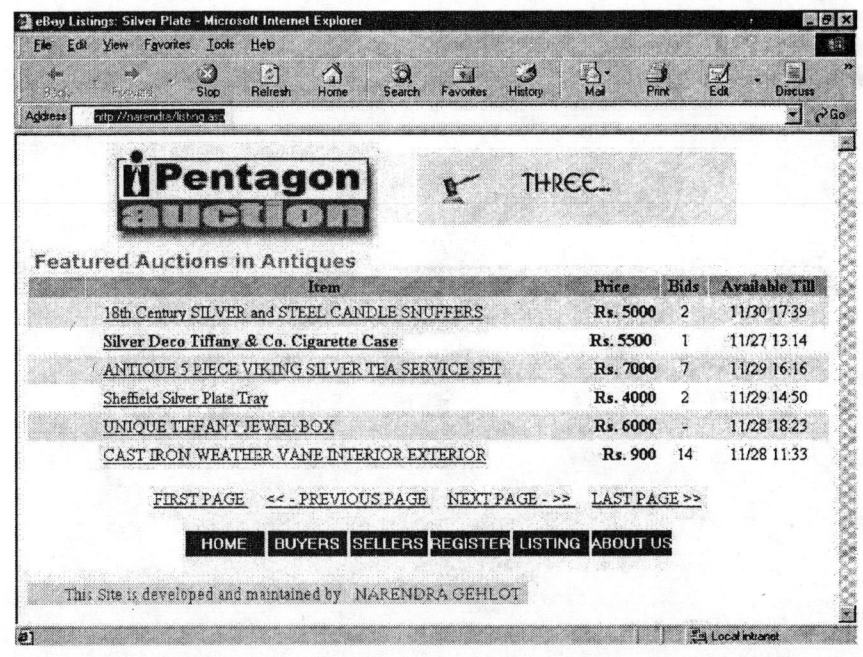

Figure B-10 Sample Item Listing posting page for myshop.com portal.

Figure B-11 Sample User Registration page for myShop.com portal.

Glossary

Architecture
An organizations hardware and software pattern, e.g. a client server architecture features servers with data and programs and remote clients usually with graphical interface capable of accessing data and running programs to process it.

Browser
A program for accessing information on the World Wide Web.

Client
The user's computer in a client-server system, contains local programs and storage.

Client-server Architecture
A computer architecture in which a number of computers in the PC to workstation class are clients of larger computer which acts as a server, the server provides data and programs for the clients and in some cases does calculations for a client.

Context Diagram
Provides an overall view of a defined system. It is made up of one process symbol, external entities and the data flow that moves between them and the system.

Data Couples
It is a unit of input/output, which can be processed.

Data Definition Language
The language used with a database management system to describe the relationships among data elements.

Data Dictionary
It is a collection of data grouped together to represent a single record.

Data Element
It is the smallest unit of meaningful data.

Data Flow
It illustrates how data moves between different components on a DFD.

Data Flow Diagram (DFD)
A diagrammatic way of showing the flow of data through a system. Using this model a process view of the system and the sub-system is laid.

Data Store
The place where one or more data elements are collectively grouped to be held as a record depicted in a DFD, is called data store.

Data Structures
Made up of data elements and other substructures.

Database Management System
Software that organizes, catalogs, stores, retrieves, and maintains data in a database.

Decision Tables
Tabular representations of conditional logic of process and the set of actions to be taken depending on each conditions being met or not.

Decision Trees
An alternative to decision table, where alternative actions that may result from different combinations of conditions are carried out.

Entity
Uniquely identified object for which information is gathered and processed.

Entity-Relationship Diagram
Entity-Relationship (E-R) diagrams are graphic illustrations used to display objects or events within a system and their relationships to one another.

Entity-Relationship Model
Entity-Relationship (E-R) model is E-R diagrams and normalized table put together.

Functional Dependency
It is functional dependency by way of which, data element identification is established in relation with its primary key.

GUI
Stands for graphical user interface, windows or other similar graphics interface for inputs-outputs.

Home Page
The first page of material an organization or individual presents on the World Wide Web.

HTTP, HTML
Hypertext transfer protocol; the protocol for transmitting HTML documents. Hypertext markup language is the language used to create hypertext documents for world wide web.

Hyperlinks
The use of reference embedded in text to allow a user to follow a topic through a documents or different documents.

Iteration
Iteration is the continued activation of a module or a set of statements within a program for as long as a stipulated condition exists.

Level-0 DFD
A Level-0 DFD is similar to combined user-level context diagram, which describes the system-wide boundaries, detailing inputs to and outputs from the system and major processes.

Level-1 DFD
A Level-1 DFD details the data flows between subsystems, which make up the whole system.

Level-2 DFD
A Level-2 DFD details the files to which the data is applied in the system, and from which data is obtained. Also each individual process is shown in detail.

Modeling
Techniques used in system analysis, design and development.

Modular Programming
The subdivision of a system and of programming requirements into small building blocks to reduce programming complexity and take advantage of common routines.

Normalization
A 3-step process for deriving data tables from set of information, which is free from redundancy.

Object oriented
A systems development and programming philosophy that views system components as objects which programs manipulates; advocates claim that object-orineted programming save development time, effort and maintenance.

Primary Key
It is a unique data element on which the other dependent data element within a record depends.

Procedural Language
A language designed to facilitate the coding of algorithms to solve a problem, for e.g. COBOL.

Process Specification
Represented in structured English, showing details of logical steps within a process.

Relational Database
A database in which data are arranged in tables; columns in the table are fields in the database and rows are records.

Relationship
Interface indicating the association between two entities.

Selection
It is a choice between two, and only two functions based on a condition.

Sequence
Set of imperative statements executed in sequence, one after another within a program.

Structure Chart
Techniques used for structured program design, which allows us to create an outline of a program by specifying modules and how the modules are connected.

Structured Decomposition
Process of decomposing context diagram into detailed DFD.

Structured English
It depicts the logical steps within a process specification. It can be used as an alternative to decision table and decision tree.

Structured Programming
A modular approach to program development that emphasizes stepwise refinement, simple control structures, and short one-entry-point/one-exit-point modules.

1NF
Abbreviation used for first normal form in process of data normalization

2NF
Abbreviation used for second normal form in process of data normalization.

3NF
Abbreviation used for third normal form in process of data normalization.

Top-down Design
Planning of a system by looking first at the major function, then at its sub-functions, and so on, until the scope and details of the system are fully understood.

World Wide Web (WWW)
A series of links among related topics among computers on the internet; requests to follow a topic through different screens are handled automatically and user does not know that he or she is moving from one computer to another.

Index

Accounts
    department, 16, 28
    receivable, 27, 33

Black box testing, 113, 128
Bottom-up design, 144

Classes, 148, 150
    Super class, 150, 158
    Subclass, 150
Class diagrams, 168
Context diagram, 4, 6, 20
    functional area, 6-11
    levels of, 7–12
    organisational-level, 11, 12, 16, 17, 127
    symbol in, 6–7
    system boundaries, 13
    user-level, 13, 126
Control Couple, 89
Credit application, 33

Data abstraction, 155
Data couples, 89
Data dictionary, 4, 42, 134
  data elements, 43, 47
  data flows, 43, 44, 47
  data stores, 7, 44
Data entry
    accuracy, 69
    forms, 62–64
    screens, 69–73
Data flow diagram (DFD), 5, 6, 18, 128
    construction, 20, 21–25
    convert context diagram to, 20
    data flow arrows, 18, 21
    invoice, 32
    levels, 22
    rules, 18
    structured decomposition, 27
    vendors, 32

Data modeling, 148
Data structures, 44, 45, 52, 53–59
Database
    attributes, 51
    design, 43, 49
    integrity controls, 49
Decision tables, 98–102
Decision tree, 94, 95–98

E-R diagrams, 3–5, 34–41
    purpose of, 34
    steps, 37
    symbols used, 34
    types of E-R relationship, 35
    many-to-many, 35, 36
    one-to-many, 35
    one-to-one, 35
Encapsulation, 151, 154, 156
Entities
    external, 18, 19, 20, 23
    relationships, 5, 61
Exercises
    context diagram, 13
    data flow arrows, 18, 21
    database, 45, 60
    E-R diagrams, 36

Forms
    guidelines for design, 68
    well designed form, 70

Graphic user interface (GUI) design, 78, 79

Information systems, 1–4
Inheritance, 154
Input form design, 67
    formats, 71
    guidelines, 68
    objectives, 68

Input screen design, 74
  Classification codes, 77
Instances of a class, 156
Inventory, 24, 25
  manage, 25, 26
  notice, 24, 25

Logical data analysis, 49
  designing database, 49
  functional dependency, 50
    scope of, 50, 51
  Process of normalisation, 51
    Composite key, 52
    first normal forms, 52, 53, 57
    non-key data element, 52
    primary key, 52
    second normal forms, 52, 54, 58
    third normal forms, 52, 55

Menu design, 81
  human computer dialogue of, 81
  menu charts, 82
    question and answer dialogue, 82
Messaging, 151
  message and message passing, 152, 157
  methods, 152, 155, 162
Methodology, 1-5
  Documentation aids, 1
  procedures, 1
  techniques, 1
  tools, 1
Modeling techniques, 4

Object, 145
  classes, 160, 166
  orientation, 147, 157
  programming, 146
Object-oriented design, 143
OOAD, 144, 145
Order
  desk, 24
  items, 54
  record, 54
  report, 53
  tracking
  store, 24
  system, 8, 10, 11, 12, 24
Output Design, 83
  internal reports, 83, 86

Polymorphism, 154, 157
Procedures, 1, 2, 3
Process
  modeling, 148
  narratives, 94
  specification, 93, 94
Program
  coding standards, 107-109
  definition language, 87
  design, 87
Purchase
  department, 16, 24

Software testing technique, 110
  acceptance testing, 112
  approach to testing, 110
  black box, 113
  final tests, 111
  plan for, 110
  techniques, 112
  white box, 113
Structure charts, 89, 91, 93
  iterative invocation, 91
Structured decomposition, 27
Structured English, 102, 105
Structured methodology elements, 209-215
Structured program design, 87, 88
  program definition language (PDL), 87
Systems
  analysis, 3-4
  design, 2
  development, 1-3, 4
  development life cycle, 1-3
    phases, 3
  scope of, 2, 3, 4

Top-down
  approach, 4, 5
  structured design, 143

User-level context diagram, 7-10
User interface, 68

Warehouse
  receiving, 8, 9, 11, 25, 26
  shipping, 8, 11, 25, 26
  white box testing, 112, 113